# The Balloon Factory

ALEXANDER FRATER has been Chief Travel Correspondent of the London *Observer* and contracted *New Yorker* writer. Two of his books, *Beyond the Blue Horizon* and *Chasing the Monsoon*, were made into major BBC television films (*The Last African Flying Boat*, based on the former, won a Bafta award for best single documentary). His most recent book, *Tales from the Torrid Zone*, was published by Picador in 2004. He lives in London.

# Alexander Frater

# The Balloon Factory

### The Story of the Men Who Built Britain's
### First Flying Machines

PICADOR

# Essex County Council Libraries

First published 2008 by Picador

First published in paperback 2009 by Picador
an imprint of Pan Macmillan Ltd
Pan Macmillan, 20 New Wharf Road, London N1 9RR
Basingstoke and Oxford
Associated companies throughout the world
www.panmacmillan.com

ISBN 978-0-330-43311-2

The publishers gratefully acknowledge Faber & Faber Ltd for permission to reproduce
an extract from *Four Quartets* by T. S. Eliot.

'A Time To Dance' by C. Day Lewis from *Noah and the Waters and Other Poems*
(© The Estate of C. Day Lewis) is reproduced by permission of PFD (www.pfd.co.uk)
on behalf of The Estate of C. Day Lewis.

The acknowledgements on page 244 constitute an extension of this copyright page.

1 3 5 7 9 8 6 4 2

A CIP catalogue record for this book is available from
the British Library.

Printed in the UK by CPI Mackays, Chatham, ME5 8TD

Visit **www.picador.com** to read more about all our books
and to buy them. You will also find features, author interviews and
news of any author events, and you can sign up for e-newsletters
so that you're always first to hear about our new releases.

*For my grandchildren,*
*Joe and Maisie*

'High spirits they had: gravity they flouted'

C. Day Lewis

# Contents

# Prologue

I first heard of Sam Cody from someone who had seen him only moments before his sudden, mysterious – and remarkably dramatic – death.

Some years ago, one unseasonably warm April evening in Brighton, I happened to be sitting in a Chinese restaurant so crowded the owner asked if I'd mind sharing my table. 'Fine by me,' I said, and the words were barely out of my mouth before the chair opposite was being dragged back by a very tall old man in a baggy unlined linen suit of a kind my father used to wear. As he sat I noted a tonsure of white stubble, a large fleshy nose networked with broken veins, bloodhound cheeks that sagged over his collar and eyes of a quite extraordinary blue. 'Kind of you,' he murmured, then picked up the menu and, as if we'd known each other years, continued, 'What do I want tonight? Fish, I think; perhaps some prawns to start. Chinese cuisine, you know, has around eighty thousand recipes; I've got about forty thousand to go.' The voice was old-school and upper-class, the sort that once read the news on the BBC Home Service (pleasantly modulated, with just a hint of a bark). He leaned towards my plate. 'What've you got there?'

'Beggar's chicken.'

He went 'Hmm', peered at the menu a moment then put it down. 'Saw the oculist today,' he carried on, conversationally. 'She thinks I need spectacles.' He directed that startling azure gaze at the Chinese owner, busy uncorking a bottle of his cheapest house red without even being asked. 'Eh, Charlie? Eighty years old, and they want to give me specs.' The owner, pouring the wine, replied, 'Maybe it's

about time, Mr Rupert. Before you kill someone in that car of yours.'
To me he said, 'Big Mercedes. And he always drive too fast.'

Mr Rupert, not best pleased – the eyes seemed to turn a frosty
sapphire – lifted his glass and took a sip. 'God, even worse than last
time.' As the owner, smiling, hurried away, he sighed. 'But I dare say
he's right.'

'Come here often?' I ventured.

'I'm just round the corner.'

An ebony walking stick was propped against his chair, its handle
an exquisitely carved ivory lion sleeping with its head on its paws. It
had once belonged, he said, to the Emperor Ch'ien Lung, a Manchu
occupant of the Dragon Throne.

'But it was actually given to me by Chiang Kai-shek.'

I stared. 'Given to you *personally*?'

'In a Shanghai hospital. I had a compound fracture of the leg.
He was carrying the stick himself, and said my need was greater than
his. And just handed it over.'

A waitress, stocky and middle-aged with pockmarked cheeks,
appeared and said, 'Hullo, Rupert.' He murmured, 'Shirley, *hullo*,'
took her hand and began discussing his meal, asking questions,
discussing options, even making her laugh.

And he did it all in Chinese.

After she'd gone I asked, 'Was that Mandarin?'

'Yes, but very bad Mandarin, I've forgotten most. They humour
me here.'

I asked how he had met the Generalissimo, but he seemed not
to hear. He kept filling his glass then downing it in a few quick
swallows, by the time the waitress arrived back the bottle was almost
half gone, yet, as she placed a dish of steamed prawns in front of him,
he spoke – this time in English – without slurring a single syllable.
'My dear, they look *fabulous*.' He picked up his chopsticks and turned
to me. 'Shirley hails from Aberdeen. That's Aberdeen, Hong Kong,
where they have the floating restaurants.'

I'd been to one of those. 'Girls paddle you out in sampans,' I said, recalling a village set around an old pirate cove and a flotilla of junks (flat-bottomed, square-bowed family dwellings) bobbing, lamp-lit, on the evening tide. 'And the fish was wonderful.'

'All caught by the Aberdeen folk. They trawl the deepest oceans. And it's the women who drive the boats on, they're the real sea dogs.' He beamed at her. 'Shirley was one, weren't you? Bestriding your quarterdeck in a Force Ten blow, yelling orders at the men. Look lively now, lads! A quick hornpipe then it's up the rigging with you!'

Affectionately she said, 'Oh, Rupert, you are so silly,' and moved off to another table.

Then, without any prompting, he began talking about the Generalissimo and his walking stick, describing at some length the way he and Chiang – also Chiang's wife – had come to know each other (of which more later), even discussing the stick's value (he'd once been offered fifteen thousand for it by a stranger in the lobby of the Adelphi Hotel, Liverpool, but he reckoned thirty would be nearer the mark). Now, having started speaking, he barely stopped, jumping from pre-war China to post-war America then abruptly back to Edwardian England, in between randomly revisiting other bits of his past, rambling up and down the twentieth century like an old bear pacing its cage.

Rupert Wood (not his real name) was born in Farnborough, Hampshire, the only child of a Treasury mandarin and a well-known landscape artist – 'Grace by name, Grace by nature' – who encouraged him, rather oddly, to take flying lessons. He had his first on his sixteenth birthday, and went solo a month later. The only boy at Harrow with a pilot's licence, he planned, after Oxford, to follow his father into Whitehall. But the admissions tutors at Balliol took one look at his academic record, '*shuddered*', and advised him to join the fire brigade.

He made a career in aviation. Grace bought him an old de Havilland biplane and he found work humping newspapers from

Liverpool to the Isle of Man. Then, wearying of these dawn sorties, he moved south and did general charters, freight mostly (once he took a giant chandelier – almost half a ton of hand-cut crystal – from London to a chateau in Antibes) but also a bit of air-ambulance work. Occasionally there were passengers too, mostly wealthy folk in a hurry, like the movie star who walked up to him at Croydon Aerodrome and said, 'Hullo. My name is Chaplin and I *must* be in Geneva by 6 p.m.'

'He gave me a hundred pounds, a fortune in those days. I was looking forward to a hugely entertaining trip – the world's funniest man sitting right beside me! – but he slept the whole way.'

Croydon's grassy, pastoral, wildflower-strewn airfield was, in the Thirties, London's gateway to the world (its acres of lambent golden buttercups said to be visible from the moon), with famous faces – Chaplin's just one of dozens – moving in and out daily. Rupert shared an office overlooking the Pyrotechnic Store. 'It was tiny,' he recalled, 'but the rent would have paid for a decent flat in Mayfair.' The monthly outlay for that, along with steadily rising fuel costs, landing fees and other overheads – also too many pilots chasing too little work – led him to decide, eventually, that aviation simply wasn't worth the candle. He was about to ask a City friend for a job – 'Death by a thousand cuts!' – when, out of the blue, his mother telephoned to ask what he knew about skywriting.

He knew that in 1910 an Englishman, John Clifford Savage, had released droplets of light engine oil into hot exhaust gases which vaporized, expanded fifty thousand times and became smoke. Also that his invention was unveiled in 1922 when startled Londoners watched him write 'DAILY MAIL' across a cloudless sky. (Later that same year an American aviator, Allen Cameron, would spell out 'HULLO USA' over New York, while 'LSMFT' – 'Lucky Strike means fine tobacco' – had them gazing upwards in Philadelphia; sales there promptly soared by sixty per cent.)

A pilot friend of Grace's, Lionel 'Shorty' Longfellow, needed help with an advertising venture. Rupert, already proficient at aerobatics, learned fast. Skywriting is, in fact, mirror-writing, and, though it's done horizontally, to anyone on the ground the words seem to stand upright and vertical. Calm air at 10,000 feet is what a skywriter seeks. His message may stay in place for an hour or more then drift across country (causing frequent traffic jams; advertisers love the close attention paid at such random moments). A skywriter's violent manoeuvrings cause a compass to spin uselessly, so headings are maintained by watching lines – such as roads – down on the ground. And altitude must always be held to within ten feet.

He wrote 'Persil', 'Cadburys' and 'Oxo' over every city on the British mainland, in time becoming engaged to Beryl, a vivacious strawberry blonde from Oxo's promotions department. Then one day he turned up for work to find that she and 'Shorty' – the affable, likeable son of an Irish earl – had cleaned out the firm's bank account and flown away. And they'd done so in Rupert's aeroplane.

Learning it had been sighted near Bremen he prepared to leave at once. But Grace urged him to let them be. He was exhausted, needed a break, a change of scene; she suggested China. There were, it seemed, family ties. An ancestor – on her side – had once been British Ambassador to the Celestial Court.

So he went and, in 1935, happened to learn from a British banker on the Shanghai Bund that skywriting was unknown in China. The banker added, 'But I suspect it would go down tremendously well, and if you could do it *in Chinese* you'd make an absolute killing.' A local doctor sold him a second-hand Gypsy Moth, the doctor's daughter, Lilly, taught him Mandarin (and welcomed him into her bed). The banker had been right. Rupert's aerial calligraphy proved to be hugely popular, people clamoured to have their marriage proposals and birth announcements posted two miles high, merchants wanted their special offers splashed up there in smoke oil, even the warlords used him for propaganda and recruitment. But the latter were a fractious lot, and when one –

whose armoury included an anti-aircraft gun – accused him of being in a rival's pay, Rupert determined to have nothing more to do with them. Yet no sooner had he made his decision than the biggest warlord of all appeared: Chiang himself, in a giant black Packard Phaeton with four armed guards riding the running boards.

'I thought, oh, God, here's trouble, but all he wanted was a birthday greeting for Soong Mei-ling, otherwise known as Madam Chiang. And luck was on my side, when the day came conditions were perfect: I went up at the crack of dawn, the rising sun made the characters glow, a high-level wind caught and enlarged them beautifully. And that night I surprised her again. For the Chinese new year I'd done special firework spectaculars, adapted the plane to drop candle bombs and shoot star rockets from wing-mounted launchers, and I did one for her birthday. She loved it! Even invited me to her party.'

By now he was also getting big commissions from overseas clients and one evening, coming in to land after doing a job for a Scots distillery, failed to notice a dog on the runway. A swerve caused the undercarriage to collapse and the plane to cartwheel. He was pulled from the wreckage with a crushed leg and, when he could finally walk, his only wish was to go home. 'But Lilly wanted to come too. She was demanding a British passport *and* a Buddhist wedding. I imagined prayer flags strung across Farnborough, with crackers going off and everyone banging gongs in church. I said no, so she broke into my safe and stole all my money – the second time a woman had done that to me! I sold my Gypsy Moth, and flew out of Hong Kong with Imperial Airways on an old Armstrong Whitworth 15 . . . '

I remembered something. 'Actually, I think it was probably a de Havilland 86.'

This made him frown. 'Pardon me?'

'The aircraft they used on the Hong Kong run.'

The frown deepened. 'Was it indeed?'

I explained that I'd written a book about Imperial and, when I told him its title,* the frown turned into a kind of reflective scowl. 'I've read that,' he admitted. 'Not so long ago, as a matter of fact, in hospital for my hip. You followed their original London to Brisbane route, which took about three weeks and stopped all over the place. Correct?'

'Absolutely.'

Then he said, wholly unexpectedly, 'It's a book my mother would have enjoyed.' I waited for him to elaborate but instead, he emptied the dregs of the wine into his glass then held up the bottle reflectively. 'It says Riocha, produce of Spain, but I know Charlie, he probably made this in his garage. Kowloon claret, you mix grape juice, molasses and raw alcohol, also some of that cactus extract Mongolians give their horses.'

He took a long swallow. 'Farnborough was quite posh in those days – the Empress Eugénie, widow of Napoleon III, France's last Emperor, lived near us. My father, Gerald, worked in Whitehall, so had to commute. He was exceptionally clever. Those Balliol dons only agreed to see me because he'd left such a mark there (though Grace said he never seemed to do any work). He died when I was two. Holiday accident, went swimming in Windermere one evening and got in the way of a carelessly driven speedboat, rising Treasury star killed by a drunken baker from East Grinstead.'

'Why would your mother have enjoyed my book?'

'Ah!' He grinned a little blearily. 'Because of her fascination with all aspects of flight. It was Ma who encouraged me. *She* put the idea in my head. Nobody in my family had ever done anything like it before, so who put the idea in hers? My belief is that it was suggested by a neighbour. You'll be familiar with the name Sam Cody, I assume?'

---

* *Beyond the Blue Horizon.* (Imperial's fleet of flimsy little fabric-winged machines – always dispatched with muttered prayers for their safe return – operated the longest, most hazardous airline services on earth, fetching up in places with names most people couldn't even pronounce. The company, eventually, became British Airways.)

It meant nothing to me. 'Sam who?'

He gave me a look that was faintly incredulous. 'And you the author of a book about aviation! He flew the first aeroplane in Britain. Actually he was Texan, known for his temper, in the Klondike someone once shot him because of a disputed claim and, though quite seriously injured, he went after the man and clubbed him senseless with a fifty-pound bag of nails. He used to wear spurs and Colt .45s in twin holsters – he could hit a spinning coin twenty feet in the air – and a big black sombrero. Cody grew his hair right down his back, and rode a giant white horse.'

I laughed. 'In *Farnborough*?'

'Well, why not? And it was there, on the 16th of October 1908, that he finally got airborne. Not for long, and the authorities denied he'd ever left the ground; they always wanted an Englishman to be first. But my mother was there. She saw it happen.' He beckoned to Shirley. 'My dear, can you perhaps tell me where this wine *really* comes from?'

She glanced around to see if Charlie was within earshot. 'Egypt,' she murmured.

He seemed unsurprised. 'Really? Well, I wouldn't mind a drop more. And some fried belly pork with red bean curd sauce, please. Oh, and a few steamed biscuits.'

She nodded and moved off. I said, 'What on earth was your mother doing?'

'Doing where?'

'Watching Cody.'

'I don't know. Perhaps waiting for him to fix her car.'

'He fixed her *car*?'

'So I believe. He was fond of her, and she . . . well, she adored him, said he once even took her up over Laffan's Plain, which would have made her, after Mrs Cody, only the second woman in Britain to have travelled in an aeroplane. Was it true? I don't know. But I do know something triggered a lifelong interest; she read everything she could on aviation, occasionally put planes in her paintings – small, but always accurately done; even one of her Royal Academy land-scapes contained a little red biplane. My father's relations with Cody, however, were evidently quite strained. He simply didn't know what to make of this gun-toting American weirdo with his long hair and waxed moustaches and goatee beard (Cody was terribly vain) *and* a wife from the English upper classes. (Lela Cody was well connected – she knew the royal family – and very beautiful.) Also, they had absolutely nothing in common: Gerald loved books while I doubt Cody ever opened a book in his life; and Cody was a mechanical genius whereas Gerald couldn't even drive. Grace did that, and always very fast.'

I asked, 'But do you actually remember Cody?'

He hesitated. 'I'm not sure, to be honest. I heard so much about him the stories may be fused in my memory. I was four when he died. Is that too early? No, it's not, because there *is* something. On August Bank Holiday Monday 1913 he took off in his new machine, a big flying boat. It was a glorious day. He had a friend with him, a well-known cricketer named Evans who played for Surrey. They passed right over our house, quite low, and Grace apparently rushed me into the garden. "Wave! Wave!" she commanded, waving like mad herself. She says he waved back. Then, just a few miles further on, the plane broke up (to this day no one knows why); people who saw it happen said Cody, dressed all in white, fell with his arms outstretched.'

'So you remember waving?'

'My memory is an odd mix of things: sunlight, perhaps a silvery plane in a blue sky, Grace's happy agitation, the smell of her perfume. And I can remember her crying when we got the news.'

Shirley returned with a new bottle of the house wine. He said, 'My dear, fetch me another glass, I want my friend to try this.'

'Just a splash,' I said.

It was a little sugary, but quite palatable. He swirled his around and sniffed deeply. '*Cairo!*' Then he swallowed. 'Strong Nile flavours here, you can taste the bilharzia. One normally gets infected through the feet. This is *much* faster.'

I wanted to know more about Grace.

'Well, it occurred to me, much later in life, that Cody's death had affected her almost as much as Gerald's did. She rarely talked about him, but when I was still a child she started taking me down to the Balloon Factory.'

This, I learned, was a cavernous, cathedral-sized structure towering over the north side of Farnborough Common, where the spirit of Cody lived on. 'It was there he built and flew Army Aeroplane No. 1, our first manned, engine-powered aircraft. All the pioneers came, Maxim, Dunne, de Havilland, Rolls, Verdon Roe, all that lot. Colonel Capper, the commandant, made sure they learned about recent developments, put their ideas forward, discussed things. Well, I say *discussed*, but actually he liked nothing better than a good argument. It was like a rather eccentric little university, with Capper as its vice-chancellor. Grace and I went to watch the planes. Nobody minded, we even got to know some of the chaps. Geoffrey de Havilland was my favourite because whenever he took off from a country field he worried about disturbing the larks. Grace fitted in with those people pretty well.'

His pork belly arrived. I ordered lychees and ice cream and, watching his arthritic fingers clumsily setting to work with the chopsticks, learned that Grace, in her fifties, had learned to fly (probably at the London Aeroplane Club in Edgware). The very day she qualified she took a friend to Dublin, later made several solo trips to the South of France. I wanted to know about his life post-China but he glossed over that. The broken leg, set by Lilly's father with a degree of incompetence that was malicious, even criminal ('The bastard never liked me'), left him with a limp and, in cold weather, severe neuropathic pain – enough for an RAF medical board, in 1939, to ground him. He spent the war producing Air Council guidance notes for air crew converting to new aircraft types.

In 1946 he married an American girl and moved to a ranch in Wyoming. There, in a barn meant for livestock, he kept a small plane, and used it to teach his son, Ben, to fly; the boy – like him – won his pilot's licence while still in his teens, aged twenty-one joined the US Navy and flew Phantom jets off carriers. A year later he was killed in Vietnam. The marriage began to drift.

'There was no acrimony. We just reached a stage where we were barely aware of each other.' When Grace died (in her sleep, aged one hundred) she left him the Farnborough house and – he hinted mischievously – the residue of an opium fortune made by his ancestor, the old envoy to Peking. Divorced, and back home again, he bought a small hotel in Cornwall, then a two-hundred-ton Cumberland herring drifter, then a Wiltshire dairy farm (he'd welcomed his guests, put to sea and milked the cows) but none sustained his interest. What he liked best was climbing into his little Cessna and exploring England.

'On fine summer days I'd be off at the crack of dawn with my binoculars, a road atlas and a thermos of coffee. Having decided on my general direction I'd then be guided entirely by impulse – or anything unexpected that hove into view: a church, a garden, a river, perhaps a place of historical interest; whatever it might be I'd go low for a closer look and, if it really took my fancy, I'd land. Around lunchtime the drill was to find a nice pub with a paddock adjoining. Or, if I'd brought a picnic, I might choose a village cricket ground. Best days of my life, probably, only stopped on my seventieth birthday. Doctor's orders.' He was looking past me with a wistful half-smile. 'Once I didn't get home for three whole months! And I came to realize, rather late in life, what a sublime country this is.'

Most of the United Kingdom's hundred and forty civil aerodromes are rural – some with names that have the steam-trumpet resonance of Victorian railway halts: Little Gransden and Little Snoring, Halfpenny Green and Farthing Corner, Hinton-in-the-Hedges and Sherburn-in-Elmet, Wombleton-Kirbymoorside and Turweston, Compton Abbas, Swinton Meadows, Swanton Morley, Old Sarum and Sleap. Indeed, some years ago, while trying to devise a new travel series for my newspaper, the *Observer*, I'd been to Compton Abbas. Wondering if it might be possible to explore rural Britain through its small planes and pastoral aerodromes I'd found myself, one sunny Sunday morning, in a spacious clubhouse thronged with enthusiasts who had spent the night under their planes – fifty or sixty light aircraft were parked along a gently sloping stretch of Dorset hillside. 'A decent sleeping bag and a wing broad enough to keep the dew off,' said a lean, scholarly looking pilot who turned out to be a retired Oxford judge. 'That's all you need.' He and his wife sat eating a late breakfast as, outside, hares chased each other across the grass strip and, beyond its perimeter, sheep grazed.

An affable young man named Miles flew me to Old Sarum, a ten-minute skip and jump across ancient green landscapes bathed in England's glorious island light. The young air-traffic controller who gave us permission to land sat yawning over a Sunday paper as his four small blond sons played around him. For various reasons the series never got done but, sitting there in Brighton, I told him about that day. Eagerly he said, 'Old Sarum, you know, was the scene of one of Cody's greatest triumphs, witnessed by Churchill and the King; both rushed to congratulate him. The King made Cody a colonel on the spot.' Idly watching him as he chattered away, face flushed with wine, an idea that had been gnawing away suddenly became blindingly obvious.

But how did you raise such a matter with someone you barely knew?

I said, 'Would you say you were very like your father?'

He looked surprised. 'I'm not like him at all. Why do you ask?'

'Dunno, just a thought.' But I persisted. 'So who were you like?'

Now I got an amused look. He broke a toothpick in four pieces and laid them end to end. 'Often, actually, I've wondered myself. But I don't think Grace was interested in sex. Hero-worship was more her style; she liked men of action, had almost schoolgirl crushes on them. Along with Cody there was Bernard Freybourg, a friend of Gerald's. He swam the Hellespont and won the VC and became a general. She adored him too.'

'Was she pretty?'

He shrugged. 'Not particularly. Stringy build, long face, a bit horsy, I suppose; big teeth – tombstones, she called them – very blue eyes, like mine. And absolutely no fashion sense; Grace always looked as if she'd been dressed by Oxfam. But, my God, she was fit, she could knock off thirty miles a day, striding across country laden down with her painting gear and getting snappy at anyone who didn't keep up. She was quite mannish in a way. Frankly, I doubt Cody would have fancied her.'

He had finished his bottle of Nile Delta Riocha, so I divided the remains of my Barossa Valley Shiraz between us. 'Would you have liked to have been Cody's son?'

'He had his own sons.'

'Even so.'

'Actually,' he said, 'I was perfectly happy just being Grace's.'

And there the matter ended.

When Shirley brought our bills he grabbed both, added the totals and said, 'That comes to £103 *precisely*.' Shirley, who knew what was coming next, gave an exasperated little sigh. 'Oh, Rupert, honestly.' He grinned. To me he said, 'Toss you for it,' then fished a 50p coin from his pocket and shoved it over. 'I'll call.'

I was, perhaps, a little drunk too; this made me laugh. 'OK.' I tossed; he called and lost. Shirley helped him up, handed him his lion stick. As she walked him to the door he turned, swaying slightly. 'You know, we were unravelling the mysteries of flight when America still belonged to the Indians, our people have been engaged in this quest for bloody centuries, it's *our* story.' He shuffled on then, turning again one final time, cried clear across the restaurant, 'Why don't you write a book about it?'

## Chapter One

# Warriors Against Gravity

Eilmer of Malmesbury

One breezy morning in AD 1010 a Benedictine monk named Eilmer (also known as Elmer, Elmerus or even – due to some medieval scribal error – Oliver) walked into St Mary's, the great Saxon-built church that once occupied the hilltop where Malmesbury's Abbey now stands. He was carrying what appeared to be a pair of sails cut from linen or silk, and connected to a central staff by a row of hinged pivots. Willow spars had been spliced to the fabric in a rough cruciform, and spliced to the spars were half a dozen sisal loops, some small enough to be hand- or footholds, others larger but puzzling to the people – pilgrims and local laity alike – who stood aside to let him pass. When he entered the bell tower and began hauling his contraption up its rickety ladders none of these casual spectators could, in their wildest dreams, have imagined what he was about to do, while any fellow-clergy who knew must have prayed for his immortal soul before hurrying outside to watch. Yet among his friends there was a feeling he might even succeed; Eilmer certainly had wild ideas, but he was a respected mathematician and astrologer, a noted scientist, 'a man of good learning'. And his courage had never been doubted.

On certain days the prevailing wind, a south-westerly, seems to gather velocity over Wiltshire's chalk uplands then make for this very spot, whistling up the slope with such force it turns umbrellas inside out. ('Carry a sheet of plywood around here during a blow,' said a contemporary Abbey steward, 'and certain principles of aerodynamics will become immediately apparent to you.')

Eilmer chose such a day. Strapping on his wings – the large sisal loops were a crude body harness – he doubtless said his own prayers and stole a last look around: far below the Avon meandered by, out to the east the Cotswolds exuded their faint, mysterious shine. He stood with arms outstretched along the lateral wooden strut waiting for a gust, when it came launched himself violently into it, plunged headlong over a city wall and down the grassy Malmesbury escarpment. Then, swooping toward the river, he ran into a rogue 'current of air', evidently some form of wind shear or turbulence which he may have tried to counter by flapping his wings.

Modern experts reckon he was airborne for fifteen seconds and came to earth a hundred and fifty feet below his launch point (the equivalent of jumping from the roof of a twelve-storey building). And while his trajectory is generally agreed upon – starting near what, these days, is Abbey Row and the Old Bell Hotel, crossing Gloucester Street and the West Gate before soaring over the declivitous Silver Jubilee Garden – there is controversy over where, crippled and in unimaginable pain, he actually ended up. St Aldhelm's Meadow, Daniel's Well or an alley named Oliver's Lane all have their supporters, yet the soft, marshy soil that undoubtedly saved his life would point to St Aldhelm's. William of Malmesbury, a historian and fellow-monk, reported that, after 'collecting the breeze upon the summit of a tower, [Eilmer] flew for more than a furlong. But agitated by the violence of the wind and a current of air . . . he fell, broke both his legs and was lame ever after.'

In later years his detractors, seeing him shuffling around on his crutches and shattered limbs, whispered it was punishment for trying to emulate the angels – witchcraft in the eyes of the Church – but they were way off the mark. He had, in fact, been trying to copy the birds, and decided his wild, ill-controlled descent was caused by ineffective steering. He told William he'd 'forgotten to provide himself with a tail' so the next set of wings would have a spreading lateral rudder (inspired by a rook's feathered rump) which would enable him to tilt and turn. The instant he was released from the Abbey's infirmary he set about building them. But they were never finished. The Abbot, fearing a further attempt would prove mortal, grounded him for life.

Malmesbury is an elegant, prosperous little town off the M4 that wears its age – it was England's first capital – with a certain ease: for example, I was unsurprised to find that two youngish, smartly dressed women talking animatedly beside its covered market cross were smoking ganja, the whiff almost masked by spicy emanations from a nearby curry house. At its heart, on the brow of that hill, is the sublime Romanesque Abbey, completed in 1180 on the site of Eilmer's great Saxon basilica. It was once even larger, boasting England's tallest spire (higher, it's claimed, than Salisbury Cathedral's), which, when blown over by a fifteenth-century storm, flattened the Abbey's entire eastern end; the collapse of the giant west tower in a gale two hundred years later demolished a good part of the nave. Only a third of the original structure still stands, yet it looks oddly complete, as if an architect forced to build skyscrapers had deliberately made the spire and tower unsound; they were supposed to fall, this dreamily beautiful ruin being what he'd intended all along.

But it remains an active parish church and, passing through the intricately carved south porch – where a female cleaner, busy mopping the floor, said, 'Welcome!' – I made for the Abbey bookshop and asked a volunteer steward if they had anything on Eilmer. 'We most certainly do,' he said, and handed me a ten-page pamphlet costing a pound. 'That contains absolutely all that is known about him.'

It carried no author's name or byline, just the legend *Malmesbury Abbey Popular History Series*. I handed him a coin. 'Any idea who wrote this?'

'I did.'

So quite by chance – 'I'm only here on Thursdays' – I met the world's leading authority on Eilmer. Ron Bartholemew, keen-eyed, grey-haired and immensely affable, with aviation in his blood (he'd worked for BOAC, then British Airways), had produced the booklet as a labour of love.

We chatted, and I asked him what kind of person he thought Eilmer had been.

'Well, clearly obsessive, probably hard to live with. But a remarkably innovative thinker. I mean, there he was in the Dark Ages working out the basic principles of aerodynamics. And, of course, quite extraordinarily brave. Incidentally, if you've got a moment, we have something quite interesting here.'

Tucked away behind the south porch was a storeroom used by the verger for redundant hassocks and hymn books. It contained a small stained-glass window showing a powerfully built, strikingly handsome young man with cornflower-blue eyes and flaxen hair. 'The only image in existence,' said Mr Bartholemew. 'And they spelled his name wrong.' (It said Elmer.) 'The fact is we haven't a clue about his appearance. The only certain thing is it was nothing like this.'

In fact, the beautiful pre-Raphaelite figure in the well-cut gentian robe (posing with a set of silvery wings and an alchemist's retort) turned out to be a twentieth-century creation, one of several windows installed in 1928 to commemorate characters from the Abbey's past. 'Is he remembered today?' I asked.

'Absolutely,' he said, adding that on certain feast days people had recreated Eilmer's feat by 'flying' down a cable strung from the Abbey roof to the churchyard (and landing near the tomb of Hannah Twynnoy, killed by a tiger in 1703). 'We had stretcher-bearers there because they featured in Eilmer's story; they were the monks who'd rushed him to the infirmary. Ours were supposed to just stand around, be part of the scenery, but some of our Eilmers descended at such a lick they had to be blue-lamped off to hospital. Now the council have stepped in. There's been a decree from Health & Safety saying we're not allowed to use humans anymore; in future, Eilmer will have to be a dummy.'

Then he recounted the saddest incident in the town's long and unique association with aviation. In 1881 Malmesbury's MP, Walter Powell, decided that since the newfangled science of aeronautics had been created in his constituency, he should learn something about it. He and a friend approached a local balloonist, who agreed to take them up. Barely had they left the ground however than, for reasons never recorded, the friend and balloonist leapt from the basket, leaving Powell on his own. Having made an accelerated solo ascent he went drifting west towards the Bristol Channel, was later glimpsed crossing the coast several thousand feet up – a tiny figure waving wildly, his anguished cries barely audible – before disappearing out to sea, never to be heard from again.

Before saying goodbye to Mr Bartholemew I put my final question: in which direction had Eilmer jumped? The cleaning lady who'd welcomed me earlier happened to be passing. 'I can show you,' she said.

I followed her out the door, through the graveyard, into Abbey Row, past the Old Bell and across Gloucester Street; the Avon, narrow and muddy, went looping by below. 'It's thought he went this way. The original tower was right behind us, so he'd have gone high over our heads, so to speak, soaring down across those gardens, almost to the river.'

I stared. 'My goodness, that's a *huge* distance.'

'More than a furlong. Ron says in the original account they used a Latin unit called the stadium, which is just over six hundred feet.'

I asked her to recommend somewhere for lunch and she nodded towards the Old Bell. 'You could try that. It's supposed to be England's oldest hotel, built I believe in 1220. The food's not bad.'

Inside I found a warren of small, cosy public rooms, and grilled fish on the menu. But the plump, solemn girl who brought it had never even heard of Eilmer.

'I am Polish,' she said.

An elderly couple seated at the next table looked up. The man, virtually bald but with remarkably shaggy red eyebrows, told me about a local doctor who always claimed that Eilmer had come down in his front garden. 'He liked to ask visitors the date and place of the world's first air crash. They'd assume it wasn't all that long ago, probably in America, and he'd say, "In fact, it happened in the year 1010 AD," then point out of his window, "*right there*."'

I laughed. 'Good to know he's still talked about.'

'Don't know about that, but he's certainly in the public domain. There's the Flying Monk football ground, home to the Malmesbury Victoria team (known as the Vics). And I believe this very hotel, the Old Bell, has a special Flying Monk weekend break.'

Later, at the reception desk, a young man admitted, guardedly, that the weekend break in question contained Flying Monk elements: a Flying Monk-inspired glider trip, followed by a scenic river walk starting at the spot where 'he fell to earth'.

'So you know that, do you? The actual place?'

He shrugged. 'Our guess is as good as anyone else's.'

Eilmer lived into his eighties and gained a reputation for prophecy. Indeed, to the ecclesiastical historian William his life was of interest not because of the flight, but because he had *twice* seen a comet believed to presage doom and desolation. Early in 1066, six hundred years before Edmond Halley was even born, the celestial body that now bears his name passed close to the earth. Eilmer witnessed it then, just as he had witnessed it during its previous passage seventy-six years earlier. Though the first had certainly heralded bad things – one being the arrival of the Danish king, Canute, on the English throne – he regarded the 1066 sighting as 'much more terrible, threatening to hurl destruction on this country'; a few months later William the Conqueror landed in Kent.

(Early in the twentieth century two other pioneer aeronauts would also display – through dreams – unusual predictive powers: Geoffrey de Havilland foresaw the sudden demise of his brother, while John William Dunne divined death and destruction on a scale so catastrophic he wrote a book about it.)

As a boy, Eilmer had certainly been inspired by Spain's Moslem bird men (their stories brought to England by returning pilgrims). In 852, over a century and a half before his own attempt, a Moor named Armen Firman donned a wing-like cloak, sprang from a tower in Cordoba and landed without incident; his was the first recorded flight in history. Just twenty-three years later Abbas Ibn Firnas, a Berber poet and fellow-Cordoban, made the second. A scholarly, eloquent, versatile man, he composed astronomical tables along with his Arabic odes (and, when the Caliph called on his subjects to beautify their city, created for Cordoba a magnificent planetarium and water clock). At some point he began weaving a mysterious feathered cape reinforced with willow wands, its purpose becoming clear in 875 when, before huge crowds, he wore it as he too jumped from a tower. Having travelled a good distance, he came down heavily and injured his back.

Today Islamic nations celebrate Ibn Firnas, not Firman, as the father of flight; Libya has put him on postage stamps, while Iraq gave his name to an airport just north of Baghdad. (In the summer of 2003 Ibn Firnas was the scene of fierce fighting between Islamic insurgents and American marines.)

After an inexplicable hiatus lasting almost four hundred years an Italian followed Eilmer. Towards the end of the fourteenth century J. B. Dante, a mathematician from Perugia, is said to have glided over Lake Trasimeno (a distance of ten kilometres) before clattering into a church steeple and breaking his leg. In 1638 Hezarfen Ahmet Celebi, a Turk, launched himself from Istanbul's hundred-and-eighty-three-foot-high Galata Tower and, touching down unscathed on the far side of the Bosporus, was received as rapturously as an astronaut back from the moon. (An emotional Sultan Murad IV, noting his name meant 'Expert in a Thousand Sciences', presented the nation's new hero with a thousand gold pieces.) Then another Italian mathematician, Francesco Lana, drew up plans for a weightless 'aerial ship' with a vacuum at its heart; in theory, if made of chimerical materials weighing next to nothing, his ship must rise inexorably. It was only built two hundred years later, by a Frenchman using welded plates of wafer-thin Prussian brass lined with tissue paper. A glittering orb of impressive size – it had a diameter of thirty-three feet – Lana's craft generated much excitement but its vacuum chamber leaked so prodigiously it never even left the workshop.

Another French enthusiast, M. Colomb, calculated that the wings needed to support a man would need to be twelve thousand seven hundred and eighty-nine feet and two inches – that's almost two and a half miles – long.

Early in the nineteenth century, Thomas Walker, a Hull portrait painter, became interested in the ability of condors to carry off fully-grown sheep (even, occasionally, small children). A condor with a ram in its talons, he reckoned, must weigh much the same as a man of 'not less than ten stone'. Put the man in a vehicle weighing two stone and you would 'only exceed the weight of the condor one-fifth part'; then fit your machine with wings one fifth longer than a condor's, and they should support the man.

But was there a direct link between weight and wingspan? A condor, fourteen thousand three hundred and thirty-six times heavier than a hummingbird, boasted a span of twelve feet. A

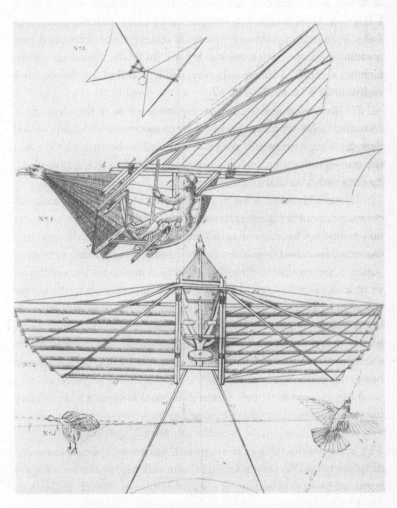

Thomas Walker's ornithopter. It never actually flew
yet Walker anticipated the day when 'thousands' of Britons
would take to the air for business or pleasure.

hummingbird had a span of three inches while a wren, three times as heavy as a hummingbird, possessed just an extra inch of wing. Logic told Walker that a man-bearing machine based on a sheep-carrying condor would need wings of fourteen and a half feet but, puzzlingly, to convey his ten-stone man, he finally settled on stubby eight-foot wings of 'compactly woven silk' that had previously been swabbed with boiled linseed oil.

His proposed aerial car – to be made of 'very thin leather' – contained a lever which, if pushed vigorously back and forth, would flap the wings; once his aeronaut had 'sufficient altitude to clear the tops of hills, trees, buildings, &c' he pressed on like an oarsman, moving with a strong and steady rhythm.

Though he never actually built his machine he was remarkably prescient about a future enhanced by machines like it: newspapers and letters to be carried at 50 mph to any part of the country, explorers launched from ships (he dreamed up the aircraft carrier) to survey a new territory 'free from the annoyance and hostilities of its rude inhabitants', and 'thousands' of Britons taking to the air for 'business or pleasure'. Demonstrating an odd touch of hippophobia he added that one consequence of so many people relying on aeroplanes would be a reduction in 'the vast number of horses kept in this kingdom'. He hoped the land needed to grow 'hay, oats and beans for the support of these quadrupeds' could be given over to producing food for the population then, on this curious note, Walker went back to painting the likenesses of Yorkshire's great and good.

The inaugural meeting of the Aëronautical (they kept that diaeresistic 'ë' for years) Society of Great Britain was held at the Kensington home of its first president, the eighth Duke of Argyll, in January 1866; its first lecture was delivered that summer at the Society of Arts in John Adam Street, beside the Strand.

It seemed to me slightly odd that people interested in science and engineering should choose such a place, but when I called there I found – painted on a pillar outside the 1770s Robert Adam building

– a more catholic prospectus: the Society of Arts was also intended to encourage 'manufacturers and commerce'. In the lobby a staircase rose past fine paintings and luminous lists of gilded names, while a fair-haired, pink-cheeked young man, obviously waiting for someone, confirmed the RSA's cultural associations – mentioned, smiling, its Polite Arts Committee's award to Edwin Landseer, aged eleven, of a medal for his depictions of animals – but added that industry and the environment had also been long-standing preoccupations. Ideas such as nationalizing the railways and digging a Channel tunnel, he said, were first floated behind the society's graceful Georgian facade – while, to counter severe deforestation in the seventeenth century, generous grants were given for tree-planting in the eighteenth. 'But they'd get involved in anything. Blue plaques were dreamed up here. And a couple of hundred years ago, when a farmer from, I think, Somerset, managed to grow opium in a barn, the RSA offered a big prize, *serious* money, to whoever could cultivate Britain's first commercial opium crop. If it could be produced by the wagonload they—'

At that moment a plump Asian girl in a green sari appeared from the street. 'Guy!' she said. 'Your carriage awaits.'

His face seemed to light up. Muttering, 'Excuse me,' he picked up his briefcase and hurried towards her.

The Society's first RSA lecturer was Francis Herbert Wenham, Bristol-based inventor of a high-pressure steam boiler robust enough to power a big ocean-going ship. He was a good-looking man with a straggly beard and intense, quizzical eyes that conveyed great intelligence and humour. In 1858 he sailed his steam-driven yacht to Egypt and, cruising up the Nile, idly began watching pelicans. Hundreds would form themselves into giant trains then, at a staggering altitude and seemingly half asleep (kept aloft by occasional lazy flaps of the wings), went '*soaring*, apparently for their own pleasure'. Soon he was noting other species too. 'The land is covered with immense numbers of blue pigeons . . .' There were wood doves and partridges, while flocks of dazzling white spoonbills raced by 'less than fifteen inches above the water's surface.' He identified, 'Eagles, vultures, kites . . . and the small, swallow-like, insectivorous hawk common to the Delta', and one day, spotting a 'magnificent' eagle on the bank, went ashore and shot it. But his lead pellets ('a charge of No. 3') simply clattered off its plumage and it flapped away, almost casually, to a 'refuge in the Lybian range'.

Gulls held a particular fascination for Wenham. Half a dozen constantly followed the yacht, one in particular coming so close he could hear the susurrant winnowing of its wings. But why did it not fall a few inches during each upward beat? A downward beat clearly gave propulsion and lift, yet the opposite produced no perceptible stagger or drop. Reason indicated that the gull's passage should be swooping and undulant, yet it 'appears to skim along a *solid* support . . . The retarding effects of gravity . . . are almost annulled, for he is gliding forward upon a *frictionless* plane.' (Wenham was much given to italics.)

He finally turned at the Second Cataract, sold his yacht to a wealthy Cairo pasha then, back in Bristol, learned from Smeaton's table of atmospheric resistances that air 'balanced' the weight of a descending surface so it could not exceed twenty-two feet per second. In 1871, pondering the relationship between pressure and velocity, he built the world's first wind tunnel and used it for an extensive study of wing shapes. Like Walker, he knew that matching weight to wing was the key. Having investigated numerous bird species, together with hornets, bees, butterflies and even beetles, he wrote his great paper on 'Aerial Locomotion'.

It stated that the curve, or camber, of a wing generated lift, and that maximum lift occurred at a wing's leading edge. Wings with a greater span (long wings) achieved better lift than those with a greater depth or chord (broad wings) even if both surface areas were the same. To reduce width he proposed that wings be 'superposed', or set one on top of the other (creating a biplane or triplane) then joined by upright supports. For maximum efficiency the pilot should lie full length and face down. And he knew that turning in the air would not be accomplished by a ship-like rudder, but by creating lift on either side. Had a gossamer-light engine been available there is a strong probability Wenham would have been the first man to fly. And it's interesting to note that, in 1899, the Smithsonian Institution sent Wilbur Wright a copy of 'Aerial Locomotion'. We know both brothers read it, that their 1903 Flyer had superimposed wings with vertical supports, that they lay full length to operate it and turned by providing lift to right or left.

In 1989, when Islamic fundamentalists began taking potshots at the boats operating Nile cruises, I was sent, by the *Observer*, to see how the boats were coping. At Luxor I joined one – five storeys high, seventy-five rooms, a rooftop pool – commanded by its cask-shaped catering manager (an ex-pastrycook) who seemed more concerned that the Islamists might cut off his liquor supply. 'This ship was built by the Yorkshire Dry Dock Company,' he assured me. 'She British as the *QE2*.'

Mornings were spent trudging around ruins, afternoons sitting on the sundeck heading upstream towards Aswan. Our course took us through a succession of deep, shadowy gardens containing people who, due to the silence of our British marine diesels, never saw us coming; black-robed women hurriedly covered their faces, gossiping men turned sharply away. Then, as the sun sank, and the river took on the colours of the sky – rose, copper, lemon and plum – we tied up for the night. (That was the moment a delicious little breeze, blowing from the north, began mixing the colours and slapping wavelets against the hull.)

Among my sundeck companions was a lanky, sharp-featured Englishman who kept pretty much to himself. In his sixties, with receding greyish hair, he took a close interest in our progress, studying the shoreline (also the birdlife) through big, old-fashioned binoculars. But I noticed him particularly because, like me, he kept scribbling in a notebook. So one day I wandered over and introduced myself. We chatted, he told me he was following in the footsteps of some crazy old Victorian who, having gone all the way up to the Second Cataract in a home-made steamboat, returned to Bristol and helped invent the aeroplane. 'Francis Herbert Wenham,' he added. 'Long-time hero of mine.' Then he spoke a bit about himself: retired aeronautical engineer, had once worked on developing Concorde (where the science, it soon became clear, possessed a beauty and certitude verging on poetry), a divorcee and – I discovered one evening, when urging whisky on him in the bar – a recovering alcoholic.

A pretty Frenchwoman in a pink bikini told us that each year millions of dragonflies, migrating down the Nile from tropical Africa, travelled so close to the water the fish jumped up and ate them. 'They are visible to the fish because their wings glitter in the sunlight,' she said. More wings! Great story! I recall my friend, at the ship's fancy-dress party, wondering, perplexedly, whether Wenham had ever heard it. (He went as Lawrence of Arabia, I as King Farouk.)

I think he was researching a book. He denied it, but I remember numerous questions about publishing, and how the system worked. I can't recall his name – and appear not to have written it down – so if ever I come across *Francis Herbert Wenham, a Biography* it may be impossible to establish if he was the author.

The first successful aero engine was created by a lace-maker from Chard in Somerset (a town thought by some historians to have been the site of Camelot). John Stringfellow was a mechanical genius who owes his place in history to William Henson, one of his Chard neighbours.

Henson, an ambitious young inventor, had moved to London where, in 1842, he dreamed up a giant passenger-carrying 'Aerial Steam Carriage' powered by two revolutionary six-bladed, steam-driven propellers. But who might actually build them and, more to the point, make them work? Stringfellow, he knew, had been

John Stringfellow

responsible for some of the finest steam looms ever to grace the West Country's lace industry. (His aquiline features, intense gaze and modishly tangled hair gave him a dreamy, poetic look, yet he spent much of his adult life smeared in grease and lubricating oil.)

He was intrigued by Henson's idea, and joined him in the capital. Together they formed the Aerial Transit Company (the world's first notional airline) and began devising a flying model. An artist's impressions of the full-sized machine that would follow had it circling the Pyramids, landing at Calcutta and crossing the Chinese coast with a Union flag flapping from its mast. Though they caused immense public excitement the press remained sceptical. So did the City's financiers. In 1848, thoroughly dispirited, Henson gave up and moved to America, ending his days, obscurely, in Newark, New Jersey.

Yet Stringfellow, now back in Somerset, persisted. Borrowing an idea from da Vinci, he devised wings with rigid fronts and flexible trailing edges, and fitted them to a toy fuselage driven by a diminutive two-stroke engine built specially for the purpose. In June 1848,

The world's first airliners were planned by John Stringfellow and
William Henson. Posters showing their Aerial Steam Carriages in
various exotic locations (this one crossing the Chinese coast)
thrilled the British public.

in a disused lace factory (witnesses included a Mr Marriatt from the
*San Francisco News Letter*), his model travelled forty yards – the world's
first powered flight.

Though he had his failures too ('There stood our protégé,' he
once wrote, 'too delicate, too fragile, too beautiful for this rough
world'), the *Scientific American* was soon noting that 'A series of experiments have recently been made beneath an immense tent in Cremorne
Gardens, London, by a Mr. Stringfellow . . . The machine excited
considerable attention and surprise by its wonderful performance. The
next expedition that is fitted out by the British Government to explore
the Niger and the country through which it winds its sluggish and

pestilential way, should employ this Mr. Stringfellow . . . to make [echoes here of Thomas Walker's imperial plans] a flying exploration untrammelled with their heels in mud or water.'

It was Francis Wenham who urged him to consider biplanes. His first, its tiny copper boiler and firebox set jewel-like in the lightest steam engine ever made (the entire machine, wrote a reporter, weighed less than a goose), performed wonderfully, while its successor thrilled the crowds at Crystal Palace. There, as a safety measure, it was sent whizzing along a high wire – spectators included an enthralled Prince of Wales – and, according to some reports, once or twice even burst free and soared away down the nave. Today this elegant little masterpiece is on display at the National Air and Space Museum in Washington DC.

Finally, like a great miniaturist determined to try his hand at a mural, Stringfellow began planning a steam plane big enough to bear him aloft. Back home in Chard he even erected a hangar in which to build it but, by now, Somerset's would-be aeronaut was in his eighties, frail and half-blind. His family, with great difficulty, talked him out of the idea and, in 1883, he died in his bed. The Wrights were mere schoolboys at the time so, had he started on his man-carrying jumbo a few years earlier, history might (down there among the apple orchards and cider presses) have been stood on its head.

When Percy Sinclair Pilcher, another West Countryman – from Bath – was a naval cadet aboard HMS *Britannia*, the blackest mark on his Admiralty record was a reprimand for breaking a teacup 'while skylarking'. He too became haunted by the idea of transcending gravity and in 1885, aged seventeen, abandoned the service and came ashore to study engineering, a decade later, as a marine-engineering lecturer at the University of Glasgow, spent his days talking about ships and his nights dreaming of 'soaring machines' capable of lifting him off the ground. His first, christened the Bat, built from Riga pine and nainsook fabric, was launched in 1895 beside the

Clyde near Cardross. After numerous attempts Pilcher urged it to a height of twenty feet and it stayed there for almost a minute. His second, the Beetle, had square-cut wings and a framework of white pine. But it proved to be cumbersome to handle – as did his third, the Gull. It was its successor, the graceful Hawk constructed at Eynsford in Kent, that drew the crowds. Built largely from bamboo – notoriously difficult stuff to work with – it was even fitted with wheels and shock absorbers. And it was buoyant as a bird.

Using techniques learned from Otto Lilienthal, the legendary German who had provided him with friendship, good advice (and even, on a visit to Berlin, flying lessons), Pilcher got his Hawk airborne for two hundred and fifty yards. The science magazine *Nature*, normally known for its icy editorial rigour, fairly babbled with excitement as, now definitely a name to be reckoned with, he prepared to take his next critical step.

Rumours began circulating that a one-horsepower internal-combustion engine, weighing just fifteen pounds, had been developed in the USA. These proved false so he decided to build his own and, moving from cramped premises in Clerkenwell to an airy new workshop in Great Peter Street, Westminster, devised a four-horsepower motor turning a five-foot propeller and weighing forty pounds; try as he might, he was unable to make a lighter version that *worked*. Knowing his Hawk was incapable of lifting such a millstone he began planning a machine that could, soon realizing that the enormous wings – and fuselage – required would, paradoxically, need an even heavier engine. The answer, had he thought to look, lay close to home in the archives of Wenham and Stringfellow, but actually came by steamer all the way from America; with admirable generosity the US pioneer Octave Chanute wrote urging him to fit *three small* wings.

It's not known how long he stayed in Great Peter Street, or why he chose a London district which now (at the apogee of Empire) included some of the most exclusive and expensive real estate on earth. And it seemed a very odd place to build an aeroplane – his

Percy Pilcher

neighbours, rich, powerful, possibly titled, would, most certainly, have complained about the noise. A number were MPs who doubtless appreciated being just a five-minute stroll from the House of Commons. But how could it have mattered to him?

I wanted to see, if possible, where he might have worked, and, one sunny winter's morning, walked from Westminster Underground station past the Palace, the Abbey and College Green (noting how the Thames seemed to lend a faint blueness to the air). The Crown Agents for the Colonies – my father had known people there – once occupied a riverside corner of Great Peter Street. Today, in its place, stood a building modishly named Four Millbank but looking comfortably conservative like, say, a St James's club. According to signs by the entrance it was home to the Crussh (sic) Juice Bar, the Atrium Restaurant ('Cocktail Happy Hour Now From 5pm – 7pm') and the studios which transmitted BBC Parliament – possibly the least-watched TV channel in Britain. An inscription said that somewhere inside Four Millbank a tree had been planted to commemorate its opening, in July 1991, by Margaret Thatcher.

A shiny chauffeur-driven Bentley Arnage (costing around £180,000) sighed to a halt and out hopped a handsome, beautifully groomed couple in their late twenties. They were plainly displeased about something. She said, 'Well, *fuck* you,' and he said, 'And fuck you too, sweetie,' then, catching my startled eye, gave me an absolutely murderous glare. I moved on, marvelling at the way London's glamorous super-rich seemed to grow younger and better-looking with every passing day.

Then, just beyond the Mary Sumner Shop ('Gifts for all occasions'), and across the road from the Royal Academy of Engineering, I found what I'd come for.

Paradoxically, in this ultra-smart thoroughfare, there stood a row of old garages – once, probably, stables – together with the brick walls and locked doors that protected the backyards of several multimillion-pound eighteenth-century houses around the corner in Cowley Street. And I knew (or know now, having used Google Earth to peer over) there was ample room in there for tools, workbenches and equipment. That, undoubtedly, was where he'd had his shed.

A few yards further on I arrived at his church. St Matthew's, designed in the 1840s by Sir George Gilbert Scott, was small, pretty, intimate, and, set a few yards back from the pavement, oddly self-effacing. Inside I found a dazzling gold-leaf high altar and reredos, and, seated at the organ keyboard, a grey-haired male teacher with a young redheaded female student. He was stolidly working his way through some atonal modern piece when, unexpectedly, she said, 'Look, why don't you try it like this?' and took over, bringing it suddenly and exhilaratingly to life. I realized, with delight, that I had it the wrong way round: she was the teacher, he the student.

Then, at the corner of Great Peter Street and Perkin's Rents, I came to the spot where, most probably, Pilcher once drank. The Speaker was cosy, wood-panelled and quiet, with a shelf of books above the bar and an eccentric display of quarrying explosives – blasting abelite and polar gelignite – mounted on the wall. Ordering a pint of Dark Destroyer from a young barman, I asked if the Speaker had been open for business late in the nineteenth century. He didn't know, but a heavyset man wolfing down a sandwich nearby said, 'This used to be the Elephant and Castle. They only called it the Speaker quite recently to attract the political crowd – plenty of them around here. But, whatever the name, I'd say there's probably been a pub on this spot for two hundred years.'

The foot of Great Peter Street produced a classic example of London's social sleight-of-hand; a few yards back it had been irredeemably posh but here, reassuringly, were the Pronto Grill, the 4 Fellas Gent's Barber and a companionable press of people. Yet I suspected Pilcher wouldn't have dallied. The smart end had too strong a pull on him – it was why he built his machine there. He *liked* the upper classes.

His triplane, assembled from seasoned pine and sailcloth, was finally completed at Stanford Hall, Lutterworth, in Leicestershire. It was the stately home of an aristocratic admirer named Braye, and Pilcher, who happened to be broke, suggested they hold a fundraiser: if Lord Braye cared to invite his rich friends along then, God willing, he would take off in front of their very noses. The prospect of the world's first powered flight rising from his manicured lawns, circling his estate's historic little haha and skimming the Avon – which went winding through his grounds – was irresistible. September 30, 1899 was chosen for the attempt, invitations sent, guests advised to bring money. They turned up in droves, the more far-sighted hoping to witness, in the dog days of that century, something momentous enough to shape and define the next.

There are two versions of what happened. Both agree on the weather: it was foul. The first version says Pilcher's engine wouldn't start, the second that the ground was so sodden he refused even to get in his plane. The crowd grew restless. By chance he had also

Pilcher's exquisitely made Hawk glider, displayed in the
Royal Scottish Museum.

brought the Hawk to Stanford Hall, and perhaps it was Lord Braye
who pointed out that if he wanted donations from these people, some
kind of show must be put on.

Two horses were hitched to the glider. They galloped off, it rose,
the tow line parted and it returned gently to earth. The crowd,
applauding wildly, demanded more. This time Pilcher reached a
height of thirty feet then, as he began demonstrating turns, a guy
wire snapped, the tail disintegrated and he came down with 'a
crash' that, according to the Hon. Adrian Verney-Cave, 'could be
heard some hundreds of yards.' Pilcher was hauled comatose from
the wreckage and rushed to the great house (renowned, ironically,
for its fine flying staircase). Aside from a broken hip he seemed to
have sustained no physical injuries yet, two days later and without
recovering consciousness, he died.

Shortly afterwards Pilcher was subsumed willy-nilly into Britain's
club for heroic failures, its members elected posthumously (a plucky

end being the price of admission) by public acclaim and a press hungry for celebrities; dead or alive, it didn't matter. Yet he never captured people's imaginations in the way that, say, Captain Scott has and, today, is remembered only by a few diehard aviation buffs. They, however, would argue that, had he lived, he'd have urged his little prop-driven machine into the sky some time that autumn – before Christmas anyway– thus beating the bleeding Yanks by four clear years.

Such claims tend to make the Yanks smile. But should they?

In 2003, at Cranfield University not far from Milton Keynes, a group of British enthusiasts set out to prove it was only the British climate that had prevented Pilcher from becoming a national treasure venerated for ever.

The lost plans of his plane were recovered and a precise replica built. Its maiden flight, with an experienced test pilot at the unbelievably rudimentary controls, was made in dead calm conditions (the Wrights needed a twenty-five-knot headwind, Force 6 on the Beaufort Scale, before they even budged). And it lasted for one minute and twenty-five seconds – a full seventy-three seconds longer than Orville, on their first hop, managed to stay aloft.

## Chapter Two

# George Cayley,
# the Yorkshire Genius

Sir George Cayley

Brompton by Sawdon lies in the Derwent valley, eight miles west of Scarborough. A national park rises to the north, to the south the Yorkshire Wolds roll off under a luminous, buttery, semi-opaque haze – or did when I was there. It takes only a minute to transit Brompton. Having passed a small war memorial and a highly rated restaurant, the Forge, you drop down by the Cayley Arms – with its mock-Tudor half-timbering – and village shop, climb away again past the School for Naughty Boys (which is how Brompton Hall tends to be described these days) and that's it. You're through.

Turn into Cayley Lane, however, and the village's hidden heart is revealed. Here are ancient trees, fine houses, deep, shadowy gardens, a stream noisy with ducks, a historic church and an air of tranquillity that is almost tangible. Its age and intimacy give the impression of a close-knit community probably unwelcoming to incomers, but Sandra and James Ineson-Parlour, who'd arrived two years earlier to open a B&B in Sawdon Lane, insisted this was not so. 'What you must do is go out and meet people,' said James, a lean, white-haired Londoner. 'Find something that interests you, and join it.' (He joined a light operatic society.) 'We felt accepted pretty much from the start.'

Sandra said, 'But the first thing you must do here is see the church, it's one of the most beautiful in Yorkshire.' She was blonde and friendly, and working locally as a supply teacher. 'Wordsworth got married to Mary Hutchinson there in 1802. Mary was keeping house for her brother Tom at Gallows Hill Farm just down the road (it's still in business) and we're told she had to set off for her wedding at the crack of dawn – well, eight in the morning, actually. Afterwards a post-chaise took her and William back to Grasmere.'

Strolling towards the church I came upon a poster announcing that Brompton's Raised Eyebrow Theatre Company would be running an Intensive Summer School in the Village Hall. ('A fun-packed and creatively vibrant week' was promised by its Artistic Director, Lizzie Patch.) Reflecting that were I to move here Ms Patch would be someone I would like to meet, and her Company something I would certainly join, I passed through All Saints' bird doors and found myself in an outer porch papered with notices: information relating to the Annual Duck Race, a single page description of Wordsworth's nuptials, an invitation 'to a pie and peas supper to celebrate the harvest', even a few lines of jolly verse: 'Never pass a little church / Always pay a visit / So when at last you're carried in / The Lord won't say / "Who is it?" '

Inside I found memorials to a family which, for generations, had headed the community and, along the way, produced an extraordinary member: Sir George Cayley. All Saints was, in effect, the Cayley family oratory: arms engraved on a flagstone commemorated Edward Cayley, the lovely stained-glass Village Window (showing a dismayed-looking batsman being clean bowled in a village cricket match) Sir Kenelm and Elizabeth, Lady Cayley, a clock Captain Sir Everard and his eldest son, Second Lieutenant F. D. E. Cayley, both killed in the Great War. Down below, in the family vault – bricked up in 1890 – lay the paterfamilias, Sir George himself. (Back in the days when he was patron of All Saints they burned, *per week*, three-quarters of a ton of coal to keep the congregation warm.) The original church, probably timber, was sacked by Vikings in 867; the present north wall dates from the thirteenth century, the tower from the fourteenth, its vice, or spiral, staircase from the fifteenth, its tenor bell from the sixteenth – since when things have been happening (chancels and vestries rebuilt, new walls put up, new bells hung, restoration work going on) – right to the present day.

Making for the Butts, a grassy common where archers used to compete, I paused to chat to a young man washing his BMW. He said the Kings of Northumbria once lived at Brompton, people had inhabited these acres for thirteen hundred years. The idea of a place occupied by forty successive generations was extraordinary, but then I recalled a sign outside the village which indicated its present inhabitants saw themselves as being, in fact, at the very forefront of the modern era – it was also, of course, the reason I had come. 'Brompton by Sawdon', it read. 'The Birthplace of Aviation'.

Sir George Cayley, born in Scarborough in 1773 but brought up here, inherited from his father a baronetcy, an immense fortune, a stately home – Brompton Hall – and great estates in Yorkshire and Lincolnshire. An engaging, handsome, high-spirited man, he was immensely interested in the natural world, and applied his formidable energy and rigorous intellect to studying and, where possible,

Brompton Hall in the 1920s. This illustration is from a brochure issued
when it became a luxury 'Country House' hotel (daily room rates starting
at about 52p). Now it's a residential school for Yorkshire boys with
'emotional and behavioural difficulties'.

improving it. His portrait shows a countryman's weathered face,
strong mouth, steady eyes, imperious nose, clean-shaven chin, high
forehead, receding blond hair, authoritative bearing, well-cut and –
but for a gaudy Paisley scarf – sober clothes. He looks politely glum,
but that's probably due to being made to sit still for so long. If the
artist had wanted a change of mood all he needed to do was hand
him, say, a shiny new clockwork toy. Then it's safe to assume (knowing what we do about George) that, face filling with interest,
murmuring *'Now what have we here?'* the true nature of the man would
have been revealed (a capacity for excitement and enthusiasm,
verging on the childlike, characterized him till his dying day).

Early promise was shown when, barely out of his teens, he devised
a way of drying out one of his vast waterlogged Lincolnshire estates
(which had lain soggy, it seems, for centuries). His ingenious 'arterial
drainage' notion worked so well that, keen to show it off and free of

charge – George certainly didn't need the money – he then reclaimed a forty-thousand-acre marsh belonging to the people next door.

George Cayley was raised just a stone's throw from the North Sea. Each winter, as its storms rattled his windows, he heard tales of the unwieldy lifeboats which, venturing out in Force 10 gales, or heading back laden with shipwreck victims, tipped over and sank. To make them tip safely back up again he devised, early in the nineteenth century, craft fitted with air-filled buoyancy tanks. George had invented the self-righting lifeboat yet, along with many of his best ideas, it would be forgotten until, years later, necessity decreed it be re-invented by someone else (and, in this particular case, made compulsory for every ocean-going vessel on earth).

Carriages rattling along England's appallingly maintained roads subjected passengers to discomfort, injury and even death (traffic accidents were commonplace); for them George invented the seat belt. Train tracks laid across public highways often resulted in passers-by getting killed, so George invented automatic signals to be placed at railway crossings (then, noting the damage that stray cattle on the line could do to a steam locomotive, invented the cow-catcher attachment that bundled them out of the way).* For accident victims who had lost limbs – and long before the word 'prosthetics' entered the language – he invented artificial hands and feet. To speed up the work on his estates he invented the caterpillar tractor (he called it the Universal Railway), the tension-spoke wheel and a steam-powered plough. By beaming light through a novel optical lens – of his own devising – he invented an instrument for testing the purity of water which, with notable success, was used to check pollution levels in the Thames. He made advances in areas as disparate as theatre design and ballistics, drew plans for an early helicopter, invented an electric motor and an early internal-combustion engine – fuelled, explosively, by gunpowder.

* So that trains needn't bother to stop he also devised a machine that scooped waiting passengers off platforms and tipped them into their carriages. It was, however, deemed impractical.

He also invented the fountain pen.

George was a founder, and first chairman, of the Regent Street Polytechnic, created in 1838 to educate the public in science and the arts (then, in 1992, reconstituted as the University of Westminster).

But he is remembered, chiefly, as the man who invented the aeroplane.

This is what the *Columbia Encyclopedia*, an American publication, says about Sir George Cayley: 'He is recognized as the founder of aerodynamics on the basis of his pioneering experiments and studies of the principles of flight. He experimented with wing design . . . formulated the concepts of vertical tail surfaces, steering rudders, rear elevators and air screws, and built the world's first glider capable of carrying a human.'

It flew in 1849, its passenger the young son of a Brompton Hall estate worker. To this day we have no idea of the boy's identity, or what became of him, or whether he ever understood the significance of what he had done. In 1853 a second, larger machine carried his coachman nine hundred feet over Brompton Dale before crash-landing in a hay meadow. Climbing breathless from the wreckage he yelled, allegedly, 'I was hired to drive, not to fly!' ('*fookin*' fly' might be closer to what was actually said) and resigned on the spot – walking out on not just his job, but an assured place in history. Today, like the boy, he remains anonymous. George, who had his own eye on posterity, left descriptions and drawings of the aircraft that carried them, but never saw fit even to mention their names. They were mere human ballast.

One July morning, exactly a century and a half later, over two thousand people watched a replica of the coachman's machine built by BAE Systems of Farnborough being carried up Brompton Vale from a lorry parked near the Brompton Village Shop. Then a helicopter appeared, circled, and put down beside a waiting group of local worthies.

Out stepped Sir Richard Branson.

The Red Arrows, he knew, were due overhead shortly. Indeed, he had been partly responsible for arranging their visit, but now he learned they were unable to take off due to fog at their home base. Since the crowd had been promised an aerial spectacular could he, at very short notice, think of an alternative? It turned out he could. A phone call was made, and a Virgin 747 bound for Heathrow from Amsterdam – on some kind of training flight, without passengers – suddenly found itself diverted to Brompton-by-Sawdon. Just moments later, however, the Red Arrows turned up after all. But when they had done their stuff and gone roaring off to their next venue (leaving the crowd noisily delighted) did Branson order his plane back on its previous course? Did he *heck*. Standing there in a grassy field, talking to the pilot on a mobile, he even issued reports on his progress: 'He's over Grimsby.' 'He can see Bridlington,' then 'He can see Scarborough,' just before everyone could see him as, in an indescribable welter of noise, he took his giant Boeing on the first of four thrillingly low passes across Brompton Vale.

Several years afterwards, in the bar of the Cayley Arms, I asked someone who had been present, 'But how low was low?' Harry, a retired businessman, stout and slightly jug-eared, said, 'Low enough to blow the washing off the line.'

'You could see the rivets,' said his pretty wife, Emma.

'And the pilot waving,' said Harry.

'That is absolutely true,' said Emma. 'We saw him waving.'

'What colour were his eyes?' I asked.

'Green,' said Harry.

'Green*ish*,' said Emma.

Harry, voice thickening with pride, added unexpectedly, 'He was a Yorkshireman.'

I laughed. 'How do you know that?'

'Richard told us.'

Then, as the 747 resumed its trip to London, it was Branson himself who became the focus of attention. A Portaloo had been placed midway up the Vale. When he stepped in nobody took much notice, but when he stepped out again people's jaws dropped. What on earth was Britain's best-known businessman doing in a frilly, lace-fronted shirt, embroidered jerkin, velvet breeches, white stockings and silver-buckled shoes? Gradually it dawned on them: Sir Richard Branson had become Sir George Cayley! The television cameras were already rolling as, wearing his period costume, he climbed into his replica plane and tested its control column. Ropes had been attached to its nose, men stood ready to pull. He raised a hand and they went charging off downhill – leaving him bumping along behind, embarrassingly earthbound; a much longer run was needed, the next attempt would have to start high on the Vale's crest. Access, however, was blocked by a fence. Branson told the fence's farmer owner, 'If you get rid of that I will have it replaced tomorrow, and pay for whatever crop you've got there, and give you two tickets to any Virgin destination of your choice. *But it must be done immediately.*'

Just then a breeze sprang up. The men with the ropes shouted, Branson scrambled back aboard – and, almost at once, was airborne. He covered thirty or forty yards at a height of about twenty feet and, afterwards, exultant, told the crowd it was the first time he had actually piloted an aeroplane. 'That,' he reported, 'was the biggest adrenalin rush I've ever had, it was just fantastic.'

Harry, sipping his gin at the Cayley Arms, said, 'It took guts, you know. People say there's no end to what he'll do for publicity, but he could have broken his neck. You'd never have got me in that bloody thing.'

'Cayley never got in it either,' I reminded him. 'It was his coachman. Branson should really have sent his chauffeur.'

I wanted to know what he was doing there at all. Harry linked it to the bitter rivalry between Virgin and British Airways. BA, it seemed, had an extravaganza planned to mark the Wright Brothers centenary. 'Then someone contacted Richard and pointed out that the hundred-and-fiftieth anniversary of aviation itself – the father of flight, by the way, being an *Englishman*—'

'From Yorkshire, actually,' said Emma.

'—fell by chance in the same year. It's probably fair to say that Richard had never heard of Cayley, but he saw that if Virgin got in first they'd get the lion's share of the media coverage. And that's what happened. We all enjoyed it, and I think he did too. He met just about everyone, he certainly met *us* – and worked that crowd like a politician, shaking hands, chatting, signing autographs. Clever chap.'

Emma told me that the appearance of his big, low-flying passenger jet had caused alarm along the coast. Voice bubbling with laughter she continued, 'People thought 9/11 was happening all over again, Manhattan first, Scarborough next, Islamic terrorists were going to crash a plane into the Esplanade Hotel. They even had to broadcast reassuring messages on local radio.'

The Revd Leonard Rivett, a Scarborough vicar who helped arrange the event, told a local journalist, 'For years I've been trying to raise the profile of George Cayley, and I hope this will do it.'

Aerial navigation had been on George's mind since he turned ten in 1783. That proved to be a remarkable year for aviation, an annus mirabilis marking the dawn of a new age, also the birth of a new nation – two of whose citizens, a hundred and twenty years later at Kitty Hawk, North Carolina, would finish the work begun by George. He, in turn, was inspired by events that began unfolding in the French provinces just as, in France's capital, the Treaty of Paris was agreed: America, finally, had been granted its independence.

In June 1783 the Montgolfier brothers, Joseph and Étienne, lit a smoky fire of straw and oily wool – chopped and shredded to make it burn better – in the marketplace of their village, Annonay. A silk bag, thirty-two feet in diameter, was suspended over the flames. It grew steadily larger then, to the astonishment of all present, *rose into the air*. In fact, the Montgolfière soared over a mile high and the French Academy of Sciences, in a state of uncharacteristic agitation, suggested that hydrogen, recently discovered by the English scientist Henry Cavendish and fifteen times lighter than air, might be substituted for smoke. The brothers, who had spent long winter evenings studying their mother's kitchen fire, used the term '*Levité*' to describe smoke's lifting power, or rather, the buoyant agent within smoke that made it capable of hoisting weights. (Later, incredulous, they would learn it was heat, not smoke, that did the trick.)

A Parisian, J.-A.-C. Charles, took note of the Academy's recommendation and, in August, from the Champ de Mars – onlookers included Benjamin Franklin and John Quincy Adams – released his Charlière, the first hydrogen-filled *ballon*. Yet, even as it came gently to earth fifteen miles away, a public debate began about the true nature of the sky. There was a belief, for example, that the earth's oxygen ran out just above the Alps (more or less at the apogee of an eagle's climb; an observer with a spyglass, noting the level at which it ventured no higher, could mark the oxygen zone's ceiling for himself). Beyond that lay a vacuum, airless yet possibly racked by dreadful invisible forces, where any human intruder risked being either choked or torn asunder.

In September of the same year, to allay these fears, Joseph Montgolfier sent up a sheep, a chicken and a duck. They returned with not a hair or feather out of place. The way was now clear for humans to try and, in November, the world's first *aéronautes*, two wealthy, flamboyant young socialites named Jean-François Pilâtre de Rozier and the Marquis d'Arlandes, boarded a balloon so gorgeously painted it resembled a small, spherical Sistine Chapel. The huge crowds waving the pair off watched them dwindle to a brilliant little exclamation mark before vanishing (surely within sight of Heaven's Gate!) altogether. They had ascended, in fact, over three thousand feet, then, concluding a notable high-speed trip, were safely back on the ground just twenty-three minutes after leaving it. (Some months afterwards Pilâtre de Rozier, forgetting that hydrogen was combustible, caused a spectacular fireball when he lit a cigar over the City of Light.)

George Cayley, at about this time, began reading the scientific periodicals – *Nicholson's Journal*, the *Philosophical Magazine* and *Mechanic's Magazine* – to which, in due course, he would become a prized contributor. And, even as he dealt with all those other subjects that kept engaging his attention, aerial locomotion remained constant, an enigma working away at the back of his mind. Balloons, he reasoned, were all very well, but they just drifted around at the mercy of the winds. Also, achieving buoyancy through the use of hot air or hydrogen was not how things were done in nature. So George turned to 'the principles of the art' that went swooping through his garden, and read 'Resistance of the Air', written in 1746 by Benjamin Robins, an eminent ballistician (or expert at determining the velocity of projectiles). Robins postulated the revolutionary theory that if you directed a stream of air against an inclined plane, the inclined plane would attempt to rise, and now George began thinking about aerial craft riding on wings. Then, at the age of twenty-six, he achieved something quite remarkable. He worked out how it was done.

The equation is inscribed, oddly, on a small silver disc. It's a diagram illustrating the four vector forces – lift, thrust, drag and weight – that govern flight while, on the obverse side – and just as revolutionary – is a picture of an aeroplane. It has a main wing, a fuselage, a cockpit, a cruciform tail with vertical and horizontal controls, and a system of revolving vanes that anticipate the propeller. The engraving is dated 1799, also the year the Rosetta Stone was discovered in Egypt; it seems not inappropriate to regard George's shiny little trinket as also being the key to a puzzle – but of such importance that its solution would change the way future generations saw the world.

George began giving his ideas physical shape and substance. He had already rejected ornithopters, or the clattering old wing-flappers supposed, by da Vinci and others, to recreate that miraculous avian combination of explosive power, sudden lift and forward motion. Power and motion would come from a motor, the wing was there to provide lift. Fixed wings, therefore, were the answer, but how exactly did they work? A century earlier Robins had invented a small portable slingshot for testing his missiles. George, adapting the idea, devised a 'whirling-arm apparatus' which, rotating at up to twenty feet a second, allowed him to observe how various experimental wing sections lashed to its tip reacted to different speeds and angles of attack. Set on a tripod, this home-made instrument recorded the earliest measurements of lift – and revealed, also, the mysterious counter-force known as drag.

In 1804, armed with his observations (pages and pages of data all recorded in that fastidious hand), he built the earliest aeroplane in history. It looked like a five-foot-long broomstick slung beneath a giant butterfly. Yet its cotton wings were set at a precise angle of six degrees, and it had an adjustable tail attached by universal joints; also it featured horizontal stabilizers, a vertical fin, and a moveable iron weight to control the centre of gravity. Launched from Brompton Dale it performed flawlessly, the world's first flight by a heavier-than-air machine.

Now he created a glider with rigid, cambered wings – made from three hundred square feet of oiled cotton – a fuselage demonstrating early streamlining and a working rudder. After repeated launchings George wrote, 'It was beautiful to see this noble white bird sail majestically from the top of a hill to any given point of the plane below it with perfect steadiness and safety, according to the set of its rudder, merely by its own weight descending in an angle of about eight degrees with the horizon.'

From time to time, when he needed books or the company of other scientists, he went to London and stayed at his town house. Hertford Street is built on a gentle incline. Today a Snappy Snaps photo shop stands at one end while the London Hilton looms massively at the other. No. 20, George's old place, is a large, handsome, four-storeyed residence standing between the Mayfair Islamic Centre and the Qatari Embassy. At No. 10, the Arab Society of Certified Accountants, blue plaques celebrate two of that house's previous owners: General John 'Gentleman Johnny' Burgoyne (who, ordered to attack New York with a large army in 1777, surrendered meekly at Saratoga), and Richard Brinsley Sheridan, playwright and politician. When a bony young man in jeans emerged from No. 20 I asked if the place belonged to him. He laughed; No. 20 was now subdivided into furnished flats, he was over from Cairo, staying in one with a friend. I pointed out Cayley's blue plaque – it called him a 'scientist and pioneer of aviation' – and he said, 'Yes, my friend tell me about him. I think he make the first flight to America.'*

Between 1809 and 1810 his great treatise on 'Aerial Navigation' appeared in *Nicholson's Journal of Natural Philosophy, Chemistry and the Arts*. The most original, and intensive, study of aerodynamics undertaken till then, it began by dispelling once and for all the myth that man could fly by flapping tied-on artificial wings. According to his calculations, the pectoral muscles that moved a bird's wings accounted for more than two-thirds of its body strength, while any combination of muscles a man might employ could never exceed even one tenth of *his* body strength; when running upstairs he might, briefly, equal the energy output 'of two of Boulton & Watt's steam horses' – yet still wouldn't generate enough to get airborne. George set about calculating how much 'propelling power' a bird (perhaps travelling 'many hundreds of miles') used, and then drew his conclusions from that.

Factors such as 'the oblique waft of the wing' led him to understand that the centre of pressure on a moving wing depended on its angle. He predicted, correctly, that when a wing was airborne a trough of low pressure would form immediately *above* it.

His measurements were extraordinarily precise. The in-flight velocity of 'a common rook', for example, was '34.5 ft. per second' while a single wing-beat, moving up and down '0.75 of a foot', would carry the bird a distance of '12.9 ft.'

---

* Just around the corner, in Hamilton Place, stand the elegant premises of the Royal Aeronautical Society – a venerable (the term 'jet propulsion' first appeared in a Society paper, dated 1867) and immensely influential organization which was actually George's idea and which, around the time of Victoria's coronation, he tried, and failed, to found; in those days aeronautics meant ballooning, and ballooning had been appropriated by dodgy showmen at country fairs. (Now the Society regards him as its spiritual mentor.)

Lightness being so critical he sought new ways of constructing a flying machine, found that 'diagonal bracing was the great principle for producing strength without accumulating weight.' Having discovered this great principle, he promptly came upon another. Sixty years before the word 'streamlining' had been coined, he wrote that 'Every lb. of direct resistance that is done away with will support 30 lbs. of additional weight without any additional power.'

He included instructions for assembling a toy helicopter – a clever idea requiring a string, a stick, two corks, eight 'wing feathers, from any bird' and a length of whalebone. Four feathers are stuck laterally around each cork, the corks are stuck on either end of the stick, the string wound once around the stick, then tied to the whalebone's extremities. The whalebone should now form a bow which, by carefully turning the feather-bladed corks *in opposite directions*, is drawn ever tighter. 'Then place . . . it upon a table . . . and with the finger on the upper cork press strong enough to prevent the string from unwinding, and taking it away suddenly, the machine will rise to the ceiling.'

In 1810 he claimed it would one day be possible to transport passengers through the air 'at 20 miles an hour'; he later revised that figure and spoke of 'ourselves and families, and . . . goods and chattels' travelling 'with a velocity of . . . 100 miles per hour'.

Naturally, to attain such an unheard of speed, a formidable 'first mover' would be needed. The hundred-pound steam engine he had in mind was, he later admitted, simply too heavy, so he turned instead to an ingenious new French motor which (not unlike the internal-combustion engine) utilized 'the sudden combustion of inflammable powders or fluids' in an enclosed space. A British version of this, fuelled by oil of tar, weighing only fifty pounds and producing eight times more thrust than his coal-fired first mover, was also available. Yet he didn't pursue it. (It's not known why; his vast wealth would indicate expense wasn't the issue.) Next he grew interested in experimental motors worked by gas and electricity then, for reasons lost to us, rejected them too.

In 1832, aged fifty-nine, he entered Parliament as Member for Scarborough. The seat was his for the taking; his great estates, his ebullience, his charm, his sociability, his money (his *millions*) had long made him a subject of fascinated local speculation. George was the most talked-about man in town and his constituents, in effect, elected him by popular acclaim – while he, interestingly, chose to represent them from the Whig benches (unlike the nation's other great landowners, who – Whigs being the forerunners of the Liberals and thus dangerously subversive – sat mostly with the Tories).

But he found a parliamentarian's duties onerous and dreary; after just two years, and having made not a single speech on the subject of aviation, he resigned and went home to Brompton Hall.

There, continuing the experiments that had preoccupied him before he left for London, he constructed a series of model gliders and took to testing them in the stairwell of Brompton Hall – but only when Sarah, his wife, was out. A brusque, authoritarian woman given to fits of sudden, barely governable rage, she had suffered painful bruising from these silent, swooping missiles, and the test ban was always reimposed the moment she got home again. George's more prescient servants knew what the trials were leading up to: that some day he'd unveil a heavier-than-air machine intended to lift one of them right off the ground.

But first, to everyone's surprise, he lifted one of their children. No objections were raised by the parents (to whom deference had been bred in the bone) and it's possible the lad himself saw nothing cruel or unusual about the idea – perhaps even thought that being the world's first fixed-wing aeronaut would make him a great celebrity. (And only just turned ten! Did shrewd old George – jingling gold sovereigns in his palm – dazzle him with this?) Lady Cayley's role is harder to fathom; her own children were probably present – indeed, it's likely the whole village turned out to watch her husband risk the life of someone else's child. Pointedly he had called this glider, with its elongated canoe-like cockpit and high, boxy wing structure, the 'Boy-Carrier' (he never bothered himself with poetic nomenclature) while making it known that a jumbo-sized version was already taking shape on the drawing board.

On the big day everything went to plan. The lad, returning safely to earth after less than a minute aloft, walked away unscathed.

The anonymous coachman who went up in the 'Man-Carrier', then crashed and mutinied in a hay meadow (no deference *there*) is remembered chiefly for having done it in 1853, exactly fifty years before the Wrights flew. According to George's meticulous drawings, his small, coracle-like conveyance was slung beneath a set of sleek, rather beautiful wings and fitted with tricycle wheels (George also invented the undercarriage). It was launched by teams of men running downhill.

An interesting rumour coming, evidently, from one of his descendants and persisting to this day, concerns the internal-combustion engine he had devised decades earlier. A new lightweight version, it alleges, was fitted to the last aeroplane he ever built. Then, with a pilot aboard – his identity, naturally, unknown – and powered 'by explosions of gunpowder discharged by a detonator', it flew successfully, its barking progress across Brompton Dale marked by trailing smoke and small eruptions of fire.

The Cayley family occupied Brompton Hall for three hundred years. When they moved out, in the 1920s, it became a luxury 'Country House' hotel (room rates starting at 10/6 a day) then much later, in a startling switch of status and function, was turned into a state-funded boarding school for boys so violent and disruptive they'd been expelled from Yorkshire's comprehensive system. While suspecting that strangers would not be welcome I phoned the office anyway, explained my reasons, and asked if it might be possible to have a quick peek inside. 'No problem at all,' said the woman who took my call. 'We've even got a little Cayley museum that might interest you.'

Brompton Hall remains a handsome, venerable stately home – bits of early fifteenth-century stonework are still in evidence – overlooking seven acres of terraced lawns and gardens (from which, according to an old brochure kept at the Cayley Arms, came the

The Brompton Hall stairwell where Cayley deployed his
famous 'whirling-arm apparatus' and – imperilling anyone using
the stairs – tested his model gliders.

'Table decorations' that had been a notable feature of the hotel). In
the vestibule where guests had been welcomed I signed in at a reception hatch, received a pass, then sat on a hard chair in an institutional
waiting room. A notice pinned to the wall said Brompton Hall was
a North Yorkshire Residential Special School for forty-six boys whose
'emotional and behavioural difficulties [had] profoundly affected
their academic performances and personal relationships.' I was
mulling this over when Lawrence Farn came hurrying through the
door. A short, stocky, likeable man with a direct manner, he was all
bustle and enthusiasm. 'I look after the museum,' he said, introduc-
ing himself. 'And I'm a teacher too.'

'Is there much interest in Cayley?' I asked, shaking hands.

'Not nearly enough.'

He showed me, first, the elegant staircase down which he had tested his models, at its foot a gleaming human-sized brass statue. 'That's Hermes, the Winged Messenger,' said Mr Farn. 'And he needs a prodigious amount of polishing, I can tell you. You know about Cayley's famous whirling arm?'

'I do.'

He pointed up to the head of the stairs. 'That's where it stood. It was worked by a pulley that dropped all the way down here to the floor.'

In a courtyard a dozen boys kicked a football around. 'And the school badge, in fact, is the Cayley coat-of-arms.' He crooked a finger. 'Lewis! Over here, please.' The boy doubled across. 'Sir?' Mr Farn, said, 'This gentleman would like to see your badge.' Lewis, a pleasant-looking lad of about twelve, with trim blond hair, showed me the Cayley emblem on his scarlet sweater. One of the boys kicked the ball to me, I kicked it on to Mr Farn and, for a few minutes, we joined in the game. Then he said, 'Care for a coffee?' and led me to the staff room, where a dozen friendly men and women were having their morning break. They offered me ginger nuts and Nescafé in a chipped mug, told me their students, aged between eight and sixteen, went home at weekends (it wasn't a Borstal) and needed a safe environment in which, firstly, to come to terms with their problems, then, secondly, to sort them out. 'We get some very disturbed kids here,' said one. Mr Farn told them of my interest in Cayley and, smiling, they recalled the Branson visit. A female teacher fetched a newspaper account of the occasion which, as an act of kindness – having first combed the school for a photocopier that worked – she photocopied for me.

The museum turned out to be a pretty little eighteenth-century summerhouse built from pale stone and set in a corner of the garden. 'This was Cayley's workshop,' said Mr Farn. He pointed to some figures roughly scratched in the stone doorway: "1810 144". 'His Man-Carrier didn't fly till 1853, but forty-three years earlier one of his model gliders was airborne for a hundred and forty-four yards; Cayley himself wrote those numbers, probably with a compass needle – to commemorate, I suspect, his first real success.'

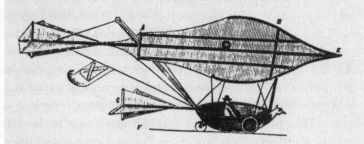

# Mechanics' Magazine,

## MUSEUM, REGISTER, JOURNAL, AND GAZETTE.

No. 1520.]     SATURDAY, SEPTEMBER 25, 1852.     [Price 3*d*., Stamped 4*d*.

Edited by J. C. Robertson, 166, Fleet-street.

### SIR GEORGE CAYLEY'S GOVERNABLE PARACHUTES.

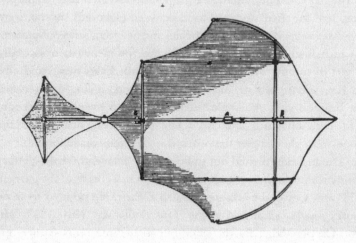

Two contemporary views of Cayley's Man-Carrier. After it crashed in
a hay meadow his enraged coachman (the behatted figure seated
in the gondola) resigned on the spot.

There wasn't, to be honest, much to see in the museum – a replica of the whirling arm, two or three model flying machines including one made by a television company which, some years earlier, had broadcast a documentary about Cayley. 'What he did was define the essence of the modern aeroplane,' Mr Farn said and, on the way out, pointed to a small wooden notice fixed to the exterior of the summerhouse. 'Scientific aeronautical experiment was pioneered from this building. Here the aeroplane was defined for the first time. Circa 1799–1855.'

Towards the end of his life, now looking oddly like the older Beethoven, he helped nurse Sarah, who had developed severe dementia. 'I'm sorry to say,' he wrote to a friend, 'that tho' in tolerable bodily health [she] has sunk into total mental imbecility.' Yet we also find him writing elatedly to M. Dupuis Delcourt, secretary of the Société Aérostatique et Météorologique de France, with news of a spectacular new Chinese top shown to him by a 'Mr Cooper of the London University'. Cooper had 'improved upon the clumsy structure of the toy' and, in George's presence, got it to a height of twenty-five feet. Back home again George improved it even further, producing a genuine super top – 'the best I have ever seen' – able to spin an extraordinary 'ninety feet into the air'.

The 'Father of Aviation' died peacefully in 1857 and, for several decades, was completely forgotten. Then, in 1909, Wilbur Wright wrote, 'About a hundred years ago, an Englishman, Sir George Cayley, carried the science of flight to a point which it had never reached before and which it scarcely reached again during the last century.'

And in 1962 the US government issued a First Day Cover for its new 8c Air Mail Coil Stamp (which showed a 747 passing over the dome of the Capitol). There, printed large on the envelope, was George's portrait with, beneath, his drawing of the elegantly streamlined 'Man-Carrier' that launched his rebellious, reluctant coachman so emphatically into history.

## Chapter Three

# The Balloon Factory

Col. James Lethbridge Brooke Templer

It was the French who first took official note of the sky, who understood it to be a strategically significant region in which, some day, trouble might foment. In 1793 they established their Military Balloon Research Centre in a graceful chateau at Chalais Meudon and filled it with some of France's finest technical minds, in 1878 they abandoned the chateau in favour of a magnificent iron-framed structure brought specially from the Paris Universal Exposition. A year later Britain cleared out an old shed at the Woolwich Arsenal, named it the Army Balloon Equipment Store and staffed it with volunteers – many of them eccentric, subversive, or simply strange – from the Royal Engineers. In 1882 they moved to Chatham, then in 1890 to Aldershot where, on the unhealthy Stanhope Lines (a recently drained swamp), Her Majesty's Balloon Factory was established. A School of Ballooning, created to ensure the nation's first air wing had a steady supply of pilots, soon followed.

Colonel James Lethbridge Brooke Templer, slim and keen-eyed, with waxed moustaches always kept precisely horizontal, was appointed Superintendent of the Factory and Commandant of the School. (Templer became well known in Aldershot for driving to the shops, on Saturday mornings, in a puffing, tall-chimneyed steam traction engine with Mrs Templer seated in a trailer behind.) He received a salary of £700 a year.

In 1895 a reporter from the *Strand Magazine* visited the School. 'The next European war,' he wrote presciently, 'will be a strange and fearful thing; everyone seems pretty sure about that.' One of the things worrying them – aside from specially bred killer flies swarming over from Germany – were balloons that rained down dynamite. Colonel Templer pointed out that all his unit had done *thus far* was spy on the enemy. A 'British war balloon', once launched, was anchored to the ground by a grapnel and a long line. Its crew, carrying field glasses, maps (two inches to the mile) and coloured pencils – red for infantry, blue for cavalry etc. – simply placed their latest markings and messages in a weighted canvas bag which went sliding down a rope to a mounted orderly, who then raced it off to the nearest general.

Next Templer talked about the way things worked: how gas was generated by mixing granulated zinc with diluted sulphuric acid, how the Factory women made his balloons from goldbeater's skin (numerous layers of the blind gut, or caecum, of an ox forming a light, immensely durable laminate), how the cord supporting the two-man cars was spun from hemp specially grown in Italy, how the cars themselves were woven by England's finest wickerwork specialists. His unit had done well in the Bechuanaland and Sudan

Expeditions, but really came of age in the Boer War. Then, observing engagements such as the Siege of Ladysmith, the Investment of Paardeberg, the Battles of Fourteen Streams, Zwarts Kop, Magersfontein and Modder River – which the British commander, Lord Methuen, would have lost but for his hawk-eyed spotters in the sky – they kept thwarting Smuts's Voortrekker snipers because the cattle-intestine balloons, being naturally self-sealing, could absorb almost endless numbers of bullets. Each hit, of course, would release a little gas and add to its ballast of lead so that, eventually, they would be forced down. But landings were made with scarcely a bump and, among the Boer marksmen, severe depression set in. For the British the success of their tethered balloons was, remarked an officer, 'one of the few bright spots in an otherwise dismal campaign.'

In 1900 several balloon companies were rushed to China for the Boxer Rebellion, but by the time they got there peace had been restored. The Cape was where they'd proved their worth, gathering such high-grade intelligence that Whitehall suddenly saw balloons playing a key role in the defence of the realm. More companies were formed and the Balloon Training School expanded. Finally the Master-General of the Ordnance himself decreed that a bigger, better Balloon Factory should be built. But where?

The old Factory at Aldershot occupied two and a half acres. Templer (who rarely appeared without a snuffbox in one hand and a large, brightly coloured handkerchief in the other) wanted at least twenty, together with a hundred and thirty wartime staff including a mounted bugler, six batmen and a shoeing smith. Assisted by C. M. Watson – later Watson Pasha of the Egyptian Army – he looked first at Weedon, a Northamptonshire village set at Britain's furthest point from the sea (balloonists hated the sea and, when aloft, refused to go anywhere near it), then at Rugby and Woking. At this juncture Colonel John Capper, an engineer who had seen service in India and Burma, was appointed Commandant of the Balloon School; Templer, however, remained in charge of the overall project – which, overnight, assumed a wholly new character when the War Office, anxiously watching Hindenburg and the Germans, decided Britain also must have an airship. The construction and testing of such a massive thing would need space. Farnborough, a small Hampshire village situated conveniently next door to Aldershot, had plenty of that, so it was decided to build it there. Templer and Capper alone seem to have realized that an airship shed must, logically, form part of any new Balloon Factory; Farnborough, therefore, was where the Factory would be.

They realized also that the sweep of land bounded by Farnborough Road to the east and the Basingstoke Canal to the south was a perfect place to locate it. Its fifteen hundred empty, level acres included Laffan's Plain, where Queen Victoria had reviewed twenty-seven thousand soldiers on her Diamond Jubilee, and Edward VII once held grand durbars. Then, falling into disuse, it had been reclaimed by the forest. In 1907, however, Aldershot's new General Officer Commanding, Sir Horace Smith-Dorrien, ordered that it be cleared. His chief engineer told him men with axes would need years to chop it down, so he called in the artillery. For six deafening hours Laffan's Plain reverberated to the sounds of sustained gunfire, crashing timber and splintering wood; when the shooting stopped not a tree remained standing. 'This,' Sir Horace later wrote, 'is how Aldershot became possessed of a flying ground.' He was now able to draw, across flat grassland, a straight, two-mile-long line reaching from the Swan Inn on Farnborough Road to Eelmore Flash on the Basingstoke Canal. And it predicted, with surprising accuracy, the path followed by Farnborough's main runway today.

In 1904 Colonel R. M. Ruck, Director of Fortifications at the War Office, began planning an 'Elongated Balloon Erecting House' on a site known as the Swan Inn Plateau. Drawings were made, and a cost of £8,900 estimated. But Ruck's superiors then allocated only £3,500 for the job, and nervously, he put it out to tender. To his intense relief the firm of Joseph Westwood & Co. Ltd said it could be done for £3,250. Having accepted their bid, however, he realized he had forgotten to include the cost of the foundations. The army had already been paid £720 to lay these, so he had, in fact, only £2,780 at his disposal. Joseph Westwood, invited to revise his estimate, said that if the War Office wanted a cheap, temporary building he could oblige for £2,760. Ruck, having agreed to both the cheap building and revised sum, then realized he had forgotten the doors. There were two, believed to be the world's largest and costing a further £300. His budget was now reduced to £2,480.

One day, when the plans were spread out on his desk, Ruck simply changed them. Using red ink he scrawled new lines, scribbled new numbers and arbitrarily reduced the width of the Erecting House by several feet. Joseph Westwood, confronted with the slimmed-down version, was invited to make yet another bid. He did so: £2,475. And, though prepared to cut certain corners – in places using thinner, 18-gauge steel sheets, for example – he refused to make any changes that could compromise safety. Ruck agreed. But then soldiers digging the foundations found that his massive new structure was about to be built on a bog. Ruck, unfazed, asked the Admiralty for its discarded gun carriages. Hewn from teak and therefore virtu-ally imperishable (probably bearing naval weaponry long before Trafalgar and the Battle of the Nile), they were also free of charge; the Admiralty was glad to be rid of them. Hundreds of carriages, comprising many tons of teak, were used to fill the excavation.

The completed building, an imposing eighty-two feet wide, a hundred and sixty feet long and seventy-two feet high, was the biggest structure for miles around. Its similarity to a cathedral was noted by Dr Percy Walker, author of a two-volume history of the Royal Aircraft Establishment, who wrote that its eastern end displayed a six-sided apse with a domed top, while the inclined bracing members – called 'rakers' – duplicated stone buttresses. (These, strongly anchored, were intended, like the ropes of a tent, to absorb tension in a strong side wind.) Then Aldershot's unsightly old Balloon Factory was disman-tled, moved two miles up the road and carefully re-erected alongside Farnborough's beautiful new Airship Shed. The Aldershot building, lower and narrower, but a hundred and fifteen feet longer, contained

The new Balloon Factory at Farnborough, with *Nulli Secundus II* passing between its giant doors (then believed to be the world's largest). This airship, unlike its predecessor, was judged not to be a success.

a Kite Room and a Skin Room where the goldbeater makers worked. (Supervised by the two Weinling brothers, Alsatian Jews wearing skull caps, several dozen women 'of the coarser sort, recruited from the rougher end of Aldershot', scraped away at their oxen's intestines. The stench would have been made worse by the proximity of the men's 'one-holer' lavatory, and a manhole set over a cesspit which, it is believed, was used by the women.) The Women's Dormitories, and Dining Room, were located not far from the Skin Room.

The two structures, along with various workshops linked by a tramway, constituted the new Balloon Factory. But when, precisely, all this happened remains unclear. Though 1905 is the generally accepted completion date, the true date, along with much else about the Balloon Factory, has fallen victim to Whitehall's code of secrecy. (In one instance it's alleged that scores of irreplaceable glass negatives relating to Farnborough's early days were neatly laid out in front of a steamroller by War Office functionaries then, under their watchful gaze, reduced to the consistency of sugar.)

What was the Factory like inside? On Belsize Road, Aldershot, there is a listed building, finished just five years afterwards, which was the Royal Engineers' Air Battalion headquarters. Its balloon store has four double doors with overlights, while period detail includes dados moulded into plaster walls, an open-well stair lined by a wreathed handrail, chimneypieces decorated with eclectic classical designs, and fireplaces around which, on winter afternoons, the aeronauts sat warming their hands and toasting buns. It's likely that the Factory also contained such welcoming and homely decorative flourishes.

Right up to the time the Post Office abolished telegrams (well into the jet age) the telegraphic address of the Royal Aircraft Establishment was 'Ballooning Farnborough'.

Meanwhile the Met Office, at South Farnborough in 1912, erected a small, functional adjunct to the Balloon Factory which possessed no architectural merit whatsoever. Yet, as the first weather outstation in Britain dedicated to gathering data exclusively for aviators, its significance was immense.

One day, a thousand feet above Newmarket, a skein of a hundred and thirty-six geese flew into a storm and perished. They had been inbound from Siberia, led (as is the goose custom) by a seasoned old gander, and, as he and his followers flapped across Berkshire, they encountered forces so unimaginable that, even *before* they tumbled to earth, their hearts had stopped beating. What happened?

The Balloon Factory aeronauts – while not having a clue – would have been unsurprised. They knew that, even when things seemed quiet on the ground, in England's volatile skies one needed to be ready for just about anything. A fellow might be tootling along enjoying the sunshine when, all at once, he was in Indian country, under attack and obliged to fight back using every trick in the book. Before

Farnborough's Met Office outstation was commissioned they got their weather by scanning clouds, holding up handkerchiefs, noting the movement of leaves, perhaps studying a lark's ascent – and wondering if any of it had a bearing on what lay across the next hill; a fine day in Frimley didn't rule out hurricanes over Hastings.

Thunderstorms, and the turbulence within them, were specially feared. At any given moment around the world eighteen hundred are active and creating a hundred lightning flashes per second – of which ten strike the ground and cause water to boil, masonry to shatter and trees to split. (In 1994 north Derbyshire recorded twenty-two thousand strikes in the course of a single day.) Ball lightning – tiny radiant spheres that drift on the wind like party balloons – can cause electrical damage (it has been known to enter houses and blow up TV sets) but is more a spectacle than a threat. Once, during choral evensong in Gloucester Cathedral, a whole galaxy floating down the Norman nave filled it with eerie luminance and caused pandemonium in the crowded pews below. Flachenblitz (rocket lightning that shoots skywards from roiling crests of cumulonimbus) is harmless, as is St Elmo's fire, which, in an open cockpit, might flicker greenly from a pilot's hair. And early aviators would have been amused by some of the stuff that has fallen from British skies: live frogs on Derby and Trowbridge, live fish on Glamorgan, premium-grade coal and fresh-cut grass (in separate showers) on Poole, pebbles on Birmingham and hazelnuts on Bristol.

Today many Britons learn to fly in places like California and Florida. They go because courses are cheaper – many schools offer all-inclusive packages – and the climate kind. Back home again, confronted by all that BLSN (blowing snow), FRA (freezing rain) and clattering GR, or hail (from the German *graupel*), the prudent will book lessons with instructors who can brief them on the surprises lurking in UK skies. These might include Gust Fronts, Line Squalls (rows of thunderstorms), Arctic Sea Smoke (known also as Steam Fog), and VLNT (violent) Valley Winds that, veering wildly, can spin a Cessna on a sixpence. There are roiling clouds with names such as Rope, Roll, Windsock and Billow (the latter three associated with hilltops, the last denoting vertical wind shear, all spelling trouble). Opacus is so thick (eight oktas, or eight eights of cover) it obscures the sun, and Cumulonimbus Mamma has vaporous bosom-shaped pendants riven with turbulence. Hills may also cause an alarming phenomenon known as Rotor Streaming.

Summer brings Black Blizzard dust spirals, and spinning vortices of hot air called Haystorms, winter FZDZ (Freezing Drizzle) and +BLSN (Heavy Blowing Snow) – all such phenomena being of interest to the meteorologists at South Farnborough's tiny weather outstation. There, helped by the pioneer aeronauts, they began devising a Morse Code glossary for those who ventured into the sky, its gnomic vocabulary used to express, back then, pretty much the same pre-departure wish list one might encounter today: a nice TLWD (Tailwind) please, and, oh Lord, if possible, NSW (No Significant Weather).

## Chapter Four

# Inherent Stability,
# or the Plane that Flew Itself

John William Dunne

John William Dunne could foretell disasters through dreams. His most famous, the nightmare that led to his book *An Experiment with Time*, occurred in the spring of 1902 when he was serving in South Africa with the 6th Mounted Infantry. One night in Lindley, an Orange Free State *dorp* recently reduced to rubble by besieging Boer forces, he dreamed he was on a volcanic island that was plainly about to erupt: vapour hissed from rocky fissures (reminding him of Krakatoa) and a terrified population crowded the beaches. Then he found himself on a neighbouring island pleading so frantically for rescue ships with its recalcitrant French authorities – 'Monsieur le Maire!' – that he was woken by the sheer force of his argument. A month later he happened to see a *Daily Telegraph* just out from London, its front page proclaiming, 'Volcano Disaster in Martinique – Town Swept Away – An Avalanche of Flames'. The news jolted him badly. What truly electrified him, however, was the date.

He had anticipated the event by precisely twenty-four hours. Not only that, but the island next door, St Lucia, had a mayor and council who spoke a French patois. Dunne began keeping a record of his dreams. Many, he would later discover, were historically accurate, while others, like the Martinique debouchment, anticipated real-life events (by no means all of them calamitous). This, eventually, led him to a world of precognitive metaphysics where past, present and future occurred simultaneously – a state likened to a book in which the pages are read not one by one but concurrently, so that its contents are absorbed in a single, dazzling, revelatory instant. Such an experience would, in theory, make us look afresh at the nature of time and reality, at existence itself (Dunne's path to enlightenment, it was argued, also proved the existence of God).

His own book, finally published in 1927, enjoyed some high-level support: C. S. Lewis and Aldous Huxley admired it, J. B. Priestley adopted its notions for his play *Time and the Conways*, while T. S. Eliot, in *Four Quartets*, wrote, 'Time present and time past /Are both perhaps present in time future, /And time future contained in time past.' Many readers, though, found it incomprehensible, while his critics claimed he had, rather crudely, blended Einstein's Theory of Relativity with Freud's ideas on the subconscious. (Yet seventy-five years later, in Gaspar Noé's 2002 movie *Irreversible*, the actress Monica Bellucci is seen lounging on the grass with a copy, and later delivers a short dissertation on Dunne's ideas. Noé's tag line was 'Le temps détruit tout', 'Time destroys everything'.)

It's probably safe to assume Dunne didn't tell Colonel John Capper, Commandant of the Balloon Factory, about his dreams;

would Capper have welcomed, into the risky, nervy world of early twentieth-century aviation, a young man who could foresee plane crashes? In South Africa Dunne had contracted enteric fever (known today as typhoid), which had left him with a faltering heart. Invalided out of the army in 1903, he called on Capper and told him about an idea he'd been mulling over since the age of thirteen: it was for an arrowhead-shaped aircraft (inspired by a Jules Verne story) that had neither tail nor rudder and would be as easy to handle as a horse. And backing came from an unexpected quarter. H. G. Wells, endlessly proclaiming the need for a 'fool-proof' British aeroplane, had befriended Dunne. He argued that the greatest danger to a British pilot came not from a putative enemy, but from his own machine which, all too often, proved bafflingly difficult to fly. The beauty of Dunne's idea was its simplicity. One's batman or groom could take control, leaving one free to study the disposition of the enemy, or look for new places in which to dig salients and trenches.

John Capper – destined to gain a knighthood and become a general – was lean-faced and square-jawed with a full, trim moustache and an unnervingly direct gaze slightly magnified by a cordless monocle clamped permanently over his left eye. Scion of a distinguished military family (the Cappers, between 1832 and 1953, won a hundred and forty-one medals) he was known for his immense zeal, uncertain temper and, crucially, his knowledge of mathematics; aeronautics, even in its infancy, needed men who understood science. A Boer War veteran himself – ex-Military Governor of Johannesburg – he struck up a close relationship with Dunne, who was charming, good-looking, well-connected in English society and a recognized expert on bats. Chewing over his V-shaped aeroplane proposal he decided that, first, a model would need to be built. Terms were agreed: 'half a guinea a day when in attendance', and all work to be done in strict secrecy.

Capper placed him in a specially curtained-off corner of the Balloon Factory and engaged a sixteen-year-old local named Gurr to assist. Gurr sat in the rafters throwing swept-wing, arrow-headed paper darts which Dunne, pacing the floor, kept passing up to him. His model, though ostensibly tailless, had, in effect, two tails – cleverly curved wingtips which, in level flight, also exerted a downward, or steadying, force. He hoped that this slightly negative incidence, combined with the backward inclination of the wings, would ensure an almost rock-like stability. (It did; he is remembered for achieving a degree of equilibrium not even the Wright Brothers could manage, also for his brilliantly innovative wing design.)

They took their finished model, made from pine and Japanese silk – its spars a set of umbrella ribs – to Caesar's Camp, a hilltop beauty spot five miles away and lying within the army's own water-catchment area. Abandoning their hansom cab at Quetta Park, he and Gurr continued on foot, trekking through a dense pine forest before arriving at the ancient fort on the summit; every aspect of this operation, as Colonel Capper had repeatedly stressed, was to be kept under wraps. That day the model performed well enough, but needed modifications. In fact, it was the first of many clandestine outings to Caesar's Camp as, working with immense patience, Dunne kept amending and refining his toy until he felt ready to construct a full-sized glider.

John Willy Dunne was, according to his brother, 'a quiet, friendly sort of man; slight in build and rather delicate from boyhood, yet possessing a certain toughness which enabled him to stand a good deal of physical and mental fatigue.' He needed that now. As the War Office kept chivvying Capper about costs, and Capper kept chivvying Dunne about progress, he nevertheless managed to build not one, but two, man-carrying prototypes which resolved the mathematics of in-flight stability so elegantly that controls became superfluous; the pilots, dangling, changed direction and altitude simply by throwing themselves this way or that. A Farnborough milliner covered the surfaces of both in black silk – which Capper then painted with white stripes. The zebra effect, he explained, was for camouflage. The First World War might still be years away, yet Capper had become obsessed with the idea that their every move was being watched by German spies. (Dunne, writing to his friend Snowden Gamble, told him the machines 'had been built piecemeal in the shops, and the parts were put together by myself and one assistant in a locked room.')

Dunne's gliders now had to be tested. Capper decreed that since Dunne's dicky heart made him unfit for flying duties, Capper himself would pilot the first. But where might this be done? Needing a place Berlin wouldn't know about, somewhere outside its ever-expanding arc of influence, he had a quiet word with Colonel Ruck, the War Office's Director of Fortifications, who then approached his shooting chum the Duke of Atholl; owner of a remote Scottish grouse moor and a fully equipped, well-trained private army, the Duke assured Ruck that both could be placed at Capper's disposal.

One evening early in 1907, Capper assembled his party at Farnborough station: Dunne and Gurr, a woodworker named Smith to effect running repairs, six sappers in mufti, and Lieutenant Lancelot Gibbs, pilot of the second glider. Lanky and heavily moustached, Gibbs had done some Alpine skiing so – reasoned the War Office grandees – must know a thing or two about plunging down gradients. (High-minded Newtonians to a man, they regarded such wilful defiance of gravity as preposterous; if people wanted to kill themselves they could more easily – and cheaply – do so by jumping off roofs.) Mrs Capper, a pretty woman soon to be befriended by the Wright Brothers, completed the group.

Tight security prevailed as they boarded their train. Waking next morning in the Highlands they were met at Blair Atholl's pine-scented station by the Marquis of Tulibardine, the Duke's eldest son, who took them to the family's vast turreted home for breakfast, then marched them up the precipitous slopes of Mount Gloag. Capper must have felt great satisfaction. Kilted Atholl Highlanders (in 1844 Queen Victoria, a house guest at Blair Castle, granted her host the extraordinary privilege of retaining his own regiment) were scouring the hillside, raking the heather and peering behind rocks; no German would last five minutes here.

Capper, wearing a fencing mask, went first. These were 'run and jump' trials to test certain aerodynamic principles, and when Dunne dropped his handkerchief Capper, giving little skips, tore off down the hill. Nothing happened. Puffing hard, he trudged back up again and fitted bicycle wheels. These took him bumping along for some distance before he was sent tumbling by a projecting rock. Mrs Capper screamed and raced towards the wreckage as, blood pouring from behind his mask, Capper rose, shouting elatedly. *There had been evidence of lift.* Now he ordered Gibbs to try. Though the soldier pedalled straight into a drystone wall he did so at such speed that, tantalizingly, he was an inch off the ground when he hit. Buoyed by this further success they hurried back to Farnborough and fitted Buchet engines: these featured twin propellers linked to flywheels rigged with bicycle spokes; you spun the wheels to get the propellers going – though it took time to figure out how you got both going together. Finally, one December afternoon in 1908, in Blair Atholl's Lower Park, Gibbs made 'a number of hops and bounds' then flew, several feet high, for forty yards.

The War Office responded to this triumph by closing down the programme. They could no longer afford, they said, to subsidize Dunne's experimental work at the Balloon Factory. If Dunne saw this calamity looming in a dream he left no record of it; and may, indeed, have been mollified when they let him keep his machines. But poor Capper, fighting to get his protégés airborne, was distraught. If the government remained so viscerally opposed to the notion of developing the aeroplane then, plainly, businessmen must take their place. He urged Tulibardine to consider the worldwide commercial possibilities of the Dunne machine. So the Marquis, together with Dunne and several aristocratic friends, set up the Blair Atholl Aeroplane Syndicate at No. 1 Queen Victoria Street, London, and vowed to continue the work. It proved to be a very sound decision.

Determined to see Blair Atholl for myself I caught a train from Inverness and shared a carriage with seven whisky-sipping old

Californians, all named Grant. They were headed for the Grant mausoleums near Carrbridge and, to frighten me, yelled their war cry: 'Stand fast, Craigellachie!' At Drumochter, today flamboyantly tricked out in rainbows, the train heaved itself over Britain's highest railway pass. There, 1,484 feet up, I saw two hog-backed hills, the Boar of Badenoch and the Sow of Atholl, and, getting off at Blair Atholl, entered a village where stands of dark trees created a green arboreal gloom. So much water dripped from them I was put in mind of a foliated sprinkler system, but the man who gave me a room at the Atholl Arms said it was merely the aftermath of a passing shower. 'Hereabouts they tend to be fast and furious.'

He knew nothing about events on Mount Gloag in 1907. 'But old Major Gordon might be able to help. He was the factor here, and a famous flying man. Has a wee plane, a Tiger Moth, which he keeps in a cowshed. His boy Andrew, he's the factor now. They're in the book.'

Calling that evening I found myself talking to three generations of Gordons. The major's son – and present factor – answered, then, as he went to fetch his father, handed me over to a child who asked about my preferred foods.

I heard footsteps approaching then, behind the murmurs, a breathy young voice whispering, '*He likes jam.*' His grandfather seemed puzzled by my request for a meeting. 'I don't know how much I can tell you,' he said. 'There's a photograph of Dunne and his friends at the castle, you should see that. But if you'd like to pop in here, shall we say, at 10 a.m. tomorrow, I'll do my best.'

Blair Atholl was a village sunk deep in its glen. Herds of afternoon cumulus grazed along the upper declivities of Beinn a' Ghio, or Mountain of the Mist, as I trudged up an avenue of limes towards the castle. One of those passing showers suddenly clattered through: tropical density, temperate chill. I passed the lushly grassed Lower Park where Lieutenant Lancelot Gibbs had been airborne for forty yards; now it was crowded with coaches and cars, and it was difficult to imagine the scene then – let alone conjure up any sense of excitement. Queuing to get into the vast, sprawling baronial mansion I overheard a Lancashire tour guide tell her party that when Mary, Queen of Scots, visited in 1564, five wolves were killed nearby. Eventually, among its collections of armour, family portraits (some by Zoffany and Landseer), Jacobite relics and Georgian toys, I came upon Major Gordon's photograph, a sepia shot of Capper standing ramrod-straight beside his skeletal machines. A large group joined me. They proved to be Germans who, to my delight, studied the old soldier and his aeroplanes with close interest. (*One even made notes.*)

Back at the Atholl Arms, slightly downcast by the cold rain and evening gloom, I went to the bar and chatted to a young Russian couple, engineers from St Petersburg who had met and married while working in Alaska.

I found myself explaining my reasons for coming to Blair Atholl.

'But this is interesting!' exclaimed the lady. 'I am descended from such a pioneer – a famous fighter pilot actually. Her name was Lidya Vladimirovna Litvak but they called her the White Rose of Stalingrad. Why? Because she had white roses painted on her Yak . . . '

'She was a small blonde with big blue eyes,' said her husband, 'though the Germans knew her more as a flying work of art, a flower study. But with guns.'

His wife said, 'She shot down twelve of them—'

'The King of England sent her a gold watch.'

'—until she herself was shot down in 1943, aged twenty-two.'

I learned that Lidya, who celebrated each kill by doing a victory roll back at base, had been among the first to volunteer for Stalin's regiments of women pilots. But when I said that the pioneers who interested me were from the early part of the twentieth century, the lady ordered another malt and talked about Nikolai Zhukovsky. He, apparently, had helped start the Imperial Russian Air Force, founded Russia's Academy of Aeronautical Engineering and made the first serious Russian contribution to the science of dropping bombs from aeroplanes. 'He also built Russia's first wind tunnel.'

I told her that the *world's* first wind tunnel* had been built by
an Englishman, Francis Herbert Wenham, back in the 1860s; she
knew about Wenham, and solemnly we drank to his memory. Over
dinner the waitress – who was Finnish – taught me several Scots
words describing specific, carefully calibrated stages of drunkenness:
bevvied, steaming, stotious and wellied. Later I drank coffee and
paged through a history of Blair Atholl, learned that in 1587 the
Earl and his Athollmen had to rescue a countess, carried off by
'Mad Colin' Campbell of Glenlyon who planned 'to use her according to his filthy appetite and lust'. A century later Simon 'the Old
Fox' Fraser of Lovat abducted and raped the Marquis of Atholl's
daughter, 'his highlanders cutting her stays with their dirks and
his pipers drowning her screams'. Steaming gently, trying to equate
such ugly deeds with this lovely place, I went to bed. A fitful wind
began blowing rain against my windows, and at about 1 a.m. a small
storm passed through, heading down the A9.

It scoured the glen which, in the morning, looked scrubbed and
shining. Major Gordon lived in a beautiful house set high on a green
hillside that dropped, sharply, to a valley floor on which a Tiger Moth
could probably land. A few low-level clouds came drifting up the glen,
one or two diffracting sunlight and sporting gaudy rainbow colours.

At eighty-two he was tall, thickset and tweed-clad, with a bone-
cracking handshake. He had a sonorous upper-class voice and an
amused look which, now, as we entered a spacious room filled with
comfortable armchairs, was turned on me. The Tiger Moth? He no
longer had any use for it. 'When a pilot reaches sixty-five he must

* Though the old Royal Aircraft Establishment, Farnborough, is now in
private hands, preservation orders have ensured that three of the nation's
finest tunnels survive there with their drive motors, lifts, hoists and six-bladed
mahogany fans intact.
For over sixty years Number One Tunnel, built in 1935, tested not only
planes and parts of planes, but also the aerodynamic qualities of Rolls-Royce
cars and the behaviour of Forestry Commission conifers in simulated storms.
The building in which it's housed, Q 121, will become a library, the tunnel itself

undergo two medicals a year; after seventy-five I simply stopped trying to impress the doctors. But I've got sons who take me up – off the lawn at the back here.'

He talked about the eldest, Andrew, who, while factoring Atholl and two other estates, flew helicopters for the army reserves. 'Long, long ago I became the first officer in the British army to qualify for helicopters, got my licence while serving with the Scottish Horse, a regiment raised by the Atholl family. Later I was on Austers, spotter planes so stable you could fly with the stick between your knees leaving both hands free for field glasses; one simply sat and contemplated the disposition of the enemy.'

'But that,' I said, 'is *exactly* what Dunne was trying for, an aircraft in which the pilot barely needed to touch the controls. It's why he and Capper came here.'

'All I know about Dunne is that he had trials in the Lower Park. And before that on Glen Tilt, a boggy bit of ground up near the top. We shoot there occasionally. They were guarded by the Atholl Hunt.'

'I thought it was the Atholl Highlanders.'

'Oh, was it? You may be right. But you know, those Austers, when I was flying them at Anzio it occurred to me that being able to spot things from the air would be a great bonus on an estate. So after the war I bought the Tiger Moth and, my goodness, I was absolutely right. It's marvellous for counting sheep – you can get close enough to see the keel marks – and, of course, deer out on the hill. I liked the

--------

– with what appears to be a twenty-four-foot extractor fan at one end – possibly a venue for receptions and book launches. In Number Two, the transonic tunnel able to separate 'swirl from air flow', Frank Whittle developed Britain's first jet aircraft, the Gloster, while McLaren used it to refine its Formula One shapes, and the UK's Olympic skiers (chins inches off the ground, supported only by the force of the gale) did their drag tests. Number Three, a low-flow turbulence tunnel in building R 52, is a thing of exceptional beauty. Built by craftsmen entirely of hardwood, it resembles the interior of an intricately made ark.

bright frosty winter mornings best, it's glorious going up to sweep the glen for stags, you can even count the points of their horns.'

He leafed through an album and produced a photograph of himself, goggled and airborne, with a golden eagle floating only yards away. 'Meeting one of these beauties always sent you home with a smile on your face. What a way to crown your day!' He added, unexpectedly pious, 'I'm the odd one out in this, of course. God gave the skies to them.' Retrieving the picture he stared at it a while. 'Ever flown around the West Coast?'

'No.'

'Ah, it's a guinea a minute up there. I own the Glen Quoich Forest and an estate near Invergarry, and I used to go over in the Moth, travelling low at a sedate seventy knots, seeing every rock and rabbit. You have to lay in your own fuel, of course, trucking it over in drums; the Highlands are badly served when it comes to landing grounds with even basic facilities. But the views are magnificent, and there are golden eagles aplenty.'

I had a train to catch but before leaving this affable, entertaining man, and curious about Geoffrey de Havilland's legendary Tiger Moth, I wondered what it was like to fly.

'Bit of a tail-dragger,' he said. 'But it's got an engine like an old Bentley.'

There had been intense rivalry between Dunne and Sam Cody to produce the first successful fixed-wing, engine-powered aeroplane in Britain. But such an ambition, openly declared, would have been considered immodest by polite society, so neither said much; either, however, could have done it. Dunne was at Blair Atholl when he learned that Cody had won – then wrecked his plane while landing. Capper, trying to lift his spirits, wrote at once, 'You will have seen the smash of Cody's machine in the papers. It flew nicely but he tried to turn too sharply and too close to the ground, with the usual result.' Capper, by fixing on the crash, and the pilot error that caused it, tried to belittle Cody's achievement. He had certainly wanted Dunne to fly first; not only was Dunne his protégé, he was also a fellow officer and member of the same upper-class military tribe (Dunne's father had been a popular army general). But envy and resentment gnawed away at Dunne for the rest of his days. The pleasant, easy-going man remembered by his brother became a dour, austere, unsmiling figure known – and usually avoided – for his utter lack of humour.

While shopping in London one day John William Dunne happened to see, in the window of Gamages, a precise toy replica of his revolutionary machine. A sales assistant, identifying it as 'The Fairey High Flying Model Aeroplane', said it had been designed by Richard Fairey, a twenty-three-year-old electrical engineer employed at the Finchley Power Station; he added, perhaps tactlessly, that it was doing exceptionally well. Dunne wrote to Fairey demanding a meeting. Fairey, who had never heard of Dunne or his patented wings, agreed. Immensely tall, with bold features and an engaging manner, he told Dunne he had perfected the design himself and, having won various gold medals for distance, decided to sell it to Gamages and make some money. (The giant Holborn store, which closed in 1972, had a toy department best known for its model trains. I once owned a Prince Charles engine, made by Bassett-Lowke, that came from Gamages.)

Then he described how he had done it. The toy-department manager, insisting that the prototype fly at least a hundred yards, took him to Hyde Park in a taxi. A lucky throw and a sudden breeze – Fairey, who loved telling the story, always called it his 'trade wind' – took it even further. Terms were discussed, Fairey cheekily naming a very large sum. The next throw took it further still. As they ran to retrieve it the excited manager yelled, 'One more flight like that, young man, and the money's yours!' But Fairey deliberately stumbled and broke the machine. (A friend, years later, remarked, 'The successful punter knows when to stop. Both men were gambling and neither lost. Fairey got his price and the store got its profit.')

Richard Fairey (second from left) with the manager and staff of
the Gamages toy department. Fairey's best-selling model plane inspired
Dunne to appoint him head of the Blair Atholl Aeroplane Syndicate.

Dunne and Fairey, hitting it off surprisingly well, began discussing aviation, even swapping ideas. Shortly afterwards, in a truly
astonishing leap of faith, Dunne invited Fairey to become manager
of the Blair Atholl Aeroplane Syndicate. Two days later, to the
dismay of virtually everyone who knew him, he left the Finchley
Power Station and embarked on what he later called 'the most
blissful period of my life'. The Syndicate now employed several
bright young Cambridge engineering graduates – one named Woofy
Dawson – who, while helping him develop 'the Dunne machine',
began initiating him into the more arcane mysteries of aeronautical
science.

Dunne's first four aircraft were built at the Balloon Factory in
1908 and 1909. The fifth, known as the Dunne D5 – designed and

financed by the Blair Atholl Syndicate, and his best yet – was constructed by the Short Brothers at Eastchurch, on the Isle of Sheppey.

In 1898 Eustace Short, a London coal merchant, bought a second-hand balloon for £30 and discovered the true meaning of happiness. He and his younger brother Oswald began making balloons – for the Indian Army – under a Battersea railway arch, and urged their older brother, Horace, to join them. Though an adventurous soul (he'd managed a silver mine in Mexico and made a miraculous escape from cannibals in the South Seas), ballooning terrified him; taken up once, he vowed never to do it again – but then became fascinated by the idea of heavier-than-air flight. In 1908 the three formed a company, moved to Muswell Manor at Leysdown on the Isle of Sheppey, and on four hundred acres of marshland, established Britain's first aircraft factory.

Their work, which began there with the production, for the Wright Brothers, of flimsy little Wright Flyers, reached its zenith three decades later when Imperial Airways introduced, on its long-haul routes, the Shorts' great multi-decked Empire flying boats. Even today these machines, built at Rochester and launched into the Medway, can induce nostalgia among those who've never even seen one. I was lucky; the day before my ninth birthday I travelled, spellbound, in an Empire from Sydney Harbour to the Fiji Islands.

Towards the end of 1909 Dunne arrived at Leysdown to meet the Shorts. Barely had they become acquainted, however, than the factory had to be moved a few miles up the road; Stonepitts Farm, at Eastchurch, was judged to be a better site for an aerodrome. Others soon joined them. By the summer of 1910 eighteen sheds stood around the new field's perimeter; Eastchurch might have been an obscure village on a marshy, thirty-six-square-mile island off the north Kent coast, yet even the Royal Aero Club, with wealthy aristocrats such as Charles Rolls and John Brabazon among its members, chose to be based there. Influential aviators like Tommy Sopwith were frequent visitors, indeed, in 1910 Sopwith won a £4,000 prize for a flight from Eastchurch to Belgium (shortly afterwards, invited home to tea by George V, he delighted the plane-mad monarch by landing on the East Terrace of Windsor Castle). In 1911, using an S.27 lent by an aviator observing an eclipse of the sun in Melanesia, the Shorts taught four navy officers to fly; soon afterwards their boss, First Lord of the Admiralty Winston Churchill, turned up demanding lessons too; he also qualified as a pilot at Eastchurch's grassy field. Now, designated (by Churchill) the nation's first Royal Naval Air Station, it was beginning to rival Farnborough and Brooklands as a centre of aviation excellence.

The Dunne D5 had a canoe-shaped capsule (with room for a passenger), a rear-mounted engine driving two pusher propellers and, of course, those elegantly curved and cambered backswept wings. At Eastchurch, on 20 December 1910, John Willy – grounded by Capper – finally got airborne. Before an audience that included Wilbur Wright he took off, lifted both hands from the controls, produced a pencil and notepad, placed the notepad on his knee and, *leaving the plane to fly itself*, calmly began to write. Those on the ground, watching astonished through spyglasses, say he wrote for 'an extended period of time', and, on returning to earth, was so preoccupied he almost collided with a windmill. But he'd made his point (also headlines around the world).

True stability.

Later a Commander Felix of the French army flew the D5 from Eastchurch to Paris and, after many 'excellent' trips around France, crashed in a field. Help would not be available for several days so, to ensure no passer-by could furtively study its 'secret features', Felix borrowed an axe and reduced Dunne's elegant machine to matchwood.

Rather more recently I chatted to a friend who had just spent a weekend on Sheppey. 'The locals call themselves Swampies,' she reported. 'It's got three prisons, a nudist beach and *hundreds* of wrecks off the coast.'

I decided it was time I paid a visit of my own. And so, on a glorious early spring day, I left the M20 and made for the lofty parabolic bridge that, soaring over the Swale channel in which

St Augustine baptized ten thousand sinners, links Sheppey to the mainland. The way to Leysdown lay through agricultural country-side occasionally disfigured by sprawling housing estates and caravan parks. I knew the Shorts had built their first planes at Muswell Manor, but didn't know where it was, or whether its past mattered to its present owners. Leysdown turned out to be a seaside holiday hamlet without a soul on its beach, or a single customer in its Fun Factory, or Dodgem Disco (£2 per car). Out towards the horizon the shadow of a giant ship stole through a bank of lucent morning mist (Sheppey was, after all, on the Thames Estuary) while an ebb tide glittering in the morning sun brought those wrecks to mind; one, the SS *Richard Montgomery*, was full of unexploded World War Two bombs. Beside Pepe's Diner two hooded, slack-jawed youths, asked about Muswell Manor, managed to look both mystified and truculent at the same time. Back at the car park I spotted a notice advertising Leysdown's attractions, and suddenly there it was: Muswell Manor Holiday Park, Birthplace of British Aviation.

At the end of a narrow coastal road I came upon a colony of caravans and caravan-sized vacation chalets with a sizeable house in the middle. It had a steeply pitched roof and big bay windows. A thickset man walked over as I drew up. 'We're closed,' he said.

I told him I was working on a book about Britain's early flyers.

He said, 'There's an exhibition about them in the bar.'

'Could I have a quick peep?'

'Sorry. You'll have to come back next month.'

Then I spotted a sign fixed to the house's wall. It sported the silhouette of a blue biplane so I hopped out of the car. 'OK if I look at that?'

He sighed.

It was a succinct history, printed on tin, of the men who had once congregated here, and what they'd achieved, and concluded with the words, 'Static/Chalets for Sale. Bar Food, Bed and Breakfast'. I became aware that someone else was present, and turned to find a good-looking, blonde-haired woman watching me intently.

A Mussel Manor photograph, taken on 4 May 1909. The three
Short Brothers stand together in the back row (Oswald, Horace and
Eustace, second, third and fourth from left) while the Wrights are
seated in front. There, from the left, are John Moore-Brabazon,
Wilbur Wright, Orville Wright and Hon. Charles Rolls.

'He's doing a book,' the man told her. To me he said, 'If you send
us a copy we could put it in the exhibition.'

She asked me, 'Have you been to Eastchurch yet?'

'No.'

'They'd have you believe it was where the Short Brothers built
the first factory, and Brabazon first flew, they're after Lottery money
for an aviation museum that would cost *millions*. So I'm in dispute
with Eastchurch.' She added, 'My name's Sharon Munns, by the
way. And he's Terry.'

Sharon spoke quietly and was very focused. Terry said hullo.

'I've seen this house in pictures,' I told her. 'With Rolls and
Dunne, the Wrights, the Shorts and everybody, all standing outside.'

'They all stayed here,' she said. 'It's a Grade II listed building that dates back to 1540. It was also known as Mussel Manor, and it was the first headquarters of the Royal Aero Club. The first pilot's licence in Britain was issued – to John Brabazon – in this very building.'

Only yards from the front door there was a big flat muddy paddock with deep puddles at one end and a caravan parked at the other. I said, 'And I guess that was the airfield.'

'Shellbeach aerodrome, the first flight made in England by an Englishman, Brabazon – though it was in a French Voison – took off from Shellbeach. Then, using a plane actually built right here at Leysdown by the Short Brothers, he flew England's first circular mile around it.'

I said I'd read that the aerodrome boasted its own golf course (Brabazon, who played daily, would years later be elected Captain of the Royal and Ancient), excellent duck- and snipe-shooting, also a fine bathing beach. The aeronauts had lived well at Mussel Manor.

'Oh, they lived like kings,' she said, and pointed to the right. 'Over there was a big barn where Rolls kept his cars, about thirty of them. When he brought the Wright Brothers down, in a Silver Ghost, he got done for speeding.' She smiled faintly. 'John Dunne stayed here. So did the Short brothers – though they put their workers up at the Coastguard's, or down at the Rose & Crown. Eustace Short died of a heart attack in 1932 while flying one of his own float planes. It was called Mussel 2, *after this house*.' She paused. 'Would you like to see the bar?'

Historic events may have given her trailer park – registered, I found later, as 'The Birthplace and Cradle of British Aviation Ltd' – an added commercial attraction, but she didn't talk like a businesswoman promoting an investment; what came through was reverence for what had once happened here, and I liked her for that. 'Yes, please,' I said. We entered a covered porch contained a notice fading from last summer ('Roast Dinners Now Being Served') and a slate plaque which, according to its inscription, had been unveiled by Lord Brabazon of Tara on 2nd of May 1999. 'That's the grandson, he came down specially. It commemorates the first flight.' A door led to the bar, now silent and shadowy, with its exhibition of contemporary photographs and press cuttings. I learned it was incomplete, still being added to, and it was she who combed the picture libraries and newspaper files. Two small girls appeared and clung to her skirts. 'Meet my granddaughters,' she said. I said hullo, then told her I was now heading for Eastchurch, and asked where I might find the old airfield. She said, 'Oh, they won't let you anywhere near that, it's been turned into a jail,' then glanced towards Leysdown's old airfield: a cow pasture owned by a local farmer, there for all to see. 'Not like ours,' she said.

Eastchurch was a pleasant, clean, spacious village set on a hill. Lunching at the Shurland Hotel, I learned the jail Sharon had referred to was Standford Hill, Britain's first open prison. My informant told me he was a warder there, asked what had brought me to Eastchurch, nodded sagely when I told him. HM Prison Service, he said, took a great interest in aviation history, and would be actively involved in the new museum. 'It'll be on the original aerodrome site, part of the prison now, but we're setting aside twenty acres for it. It's going to be fantastic, with a conference centre, a cinema, a restaurant and, of course, lots of displays celebrating the first flights. Tourists'll come by the thousand.'

'What about Muswell Manor?' I asked. 'Where the first flights actually took place?'

He shrugged uninterestedly. 'No infrastructure there, they couldn't cope. By the way, have you seen Eastchurch's memorial to the Pioneer Airmen? Just across the road? It was erected back in the Fifties and, of course, we supplied the labour, oh yes, HM Prison Commissioners showing willing again.'

The memorial, on the corner of Church Road and the High Street, was actually a bus stop. Several stone panels, carved in low relief, showed aeroplanes of the period, along with a flying helmet, navigational instruments and Zeus brandishing a thunderbolt; in front of these a row of hardwood seats had been installed. I've known bus stops boasting better views, but never one with its own dedicated committee of patrons – Churchill and Brabazon among them – or devised with such artistry and care. Reflecting that here you really wouldn't mind how late the No. 37 was running I crossed the road to an imposing fifteenth-century church, its most important window inscribed, 'To the Glory of God and in memory of Charles Stewart Rolls and Cecil Stanley Grace, Aviators July 1910.' They were the first Englishmen to die in plane crashes.

Given the nature of what John Dunne was doing, it's surprising he didn't end up the same way. After his triumphant flight at Eastchurch, he and Fairey built, in the course of the next three years, five more aircraft, always seeking to refine and improve the Dunne marque. (Who did the actual building, however, remains a mystery; he seems to have had no further dealings with the Shorts.) D10, the last of the batch, powered by 80 hp French Gnome rotaries, was flown hands-off from Farnborough to Salisbury Plain and – even more remarkable – could make perfect circles unassisted. Yet the Royal Flying Corps bought only two. Their pilots weren't interested in stability; power, agility and, above all, speed were what they wanted. In 1913 Dunne suffered a complete mental and physical breakdown. Doctors blamed overwork, he quit the business and the Syndicate sold US manufacturing rights to the owner of a Marblehead, Massachusetts boatyard named W. Starling Burgess.

In 1914 Burgess, a builder of racing yachts, produced a Dunne seaplane with a broad wooden float under the central nacelle, and pontoons at either wingtip. Handmade by craftsmen, pretty to look at, easy to fly and adored by the press, the Burgess-Dunne Hydro became America's most fashionable aeroplane. If you had the money they could even build you a floating hangar to keep it in, and America's tycoons (Vincent Astor and Harold Payne Whitney among them) rushed to place their orders. So did the less glamorous rich – a wealthy New York doctor, for example, used his Hydro to make house calls around the Adirondack lakes. The army bought one and the navy five, in 1915 Lieutenant Patrick Bellinger took his to 10,000 feet and set a new altitude record for seaplanes; another navy Hydro sported a lavender and green camouflage so gorgeously decorative that, instead of fading into the background, it was admired everywhere as a work of art. The Hydro was succeeded by the Burgess Dunne Sportsman's Seaplane, after that by a wartime reconnaissance version which, thanks to its inherent stability, enabled a pilot – or so the company claimed – 'to fly long distances without fatigue and make observations at his leisure'. One of the latter may even have been sold to Russia.

There's a curious little footnote to all this. Early in the war the Canadian government spent $5,000 on a second-hand Hydro for its newly founded Aviation Corps, then manned by two officers and a mechanic. Shipped to Britain for some reason, the Hydro broke up in a storm. Though no Canadian ever flew it, it was, in fact, the Canadian Air Force's first aeroplane, and a 1999 Canadian postage stamp celebrated this fact. 'Air Forces *Forces Aériennes* Burgess-Dunne' it said, and there was an artist's impression – painted a fetching yellow – of John Willy's supremely elegant little *avion*.

Burgess was survived by his daughter Tasha Tudor, a well-known writer and illustrator of children's books.

Dunne wrote a children's book called *The Jumping Lions of Borneo*.

He also wrote *Sunshine and the Dry Fly*, about fishing along the southern chalk streams of England. He had patented a new type of fly which, tied from coloured silks and using a white painted hook, was supposed, *when the sun shone*, to display a brilliant translucency that made it irresistible to trout. Insects 'with wings varying from crinkled pewter to the tint of Sheffield plate worn thin, and with plain, monochromatic bodies varying from the palest honey to the darkest amber' had inspired him – but did his flies work? While critics said they lost their translucency under even moderate cloud, flies tied according to his recipe were still being sold in specialist shops as late as 1966.

The book came out in 1924.

Sir Richard Fairey with his son and daughter-in-law.

In 1949 he died, regarding himself as a failure right to the end. However the young model-maker whose career he launched went on to win a knighthood, earn a fortune and become a glamorous household name; it was Fairey Swordfish aircraft – cumbersome, single-engined, ninety-knot biplanes known as 'Stringbags' – that disabled the *Bismarck* with two direct torpedo hits.

After Dunne's breakdown Richard had resigned from the Syndicate and founded the Fairey Aviation Company, in 1929 moved it from Northolt to a hundred and fifty acres of gorse-covered heath between the Bath Road and the Great South West Road – an area of west London once roamed by wolves. He gave it a level grass surface, built a large hangar on it and called it the Great West Aerodrome. In 1944 the government, liking the site, stepped in with a compulsory purchase order, and took it over. They named Fairey's aerodrome Heathrow.

\*

Back on the train again, trundling south from Blair Atholl, I pondered the link between the tranquil Scots village that gave its name to the Syndicate, and the world's busiest international airport – with a secondary link leading, via Dunne, back to the Balloon Factory. I imagined John Willy and his friends launching their weird little planes from that tussocky Highland hillside, then the awed disbelief with which they would have viewed Heathrow today. Yet it was, in a real sense, where they were always headed.

## Chapter Five

# The Chronic Inventor

Hiram Stevens Maxim with the gun that
made him rich.

Sir George Cayley died in the year that Hiram Stevens Maxim, aged
seventeen, devised his miraculous mousetrap. (Maxim's description
of himself, years later, as 'a chronic inventor' would doubtless have
brought a wry smile to the endlessly innovative Sir George.) In 1904,
at Blackpool's Pleasure Beach, he created 'Sir Hiram Maxim's
Captive Flying Machine', a sixty-two-foot-high steel pole supporting ten steel arms, from which hung carriages that, when spinning,

created the illusion of being airborne. It's still there, still working, and as popular as ever. The old metal casting remains in place, as does the original machinery: two electric motors (by Lister Brothers of Dursley), a pair of twelve-foot-diameter wheels and bevel gears that turn the crown wheel on the centre column.

Some time ago the *Observer* sent me to Blackpool to write about the place prior to a Labour Party Conference. A year earlier, when the Tories came, it was rumoured that all manhole covers had been welded down, that the Olympic-sized Derby Pool would be drained and turned into a makeshift mortuary, that three hundred body bags had been delivered for delegates staying at the Imperial Hotel and, most fanciful of all, that an IRA trawler was going to sail in from the Irish Sea and fire Exocet missiles at Mrs Thatcher.

Labour were actually looked upon with favour in Blackpool. They spent more money (though nothing like the profligate, hard-drinking Young Farmers, whose own conference was a high point of the local trading year) and swallowed more champagne, but were not nearly as controversial. The previous year, I knew, had seen a real sense of anticipated danger, but now everything seemed back to normal (gangs of boisterous vacationing Lancashire girls piling onto the trams wearing rainbow glitter wigs, the Tower Ballroom's six organists toiling away, in shifts, at their Wurlitzers). It was hard to find anyone who was even remotely interested in the Party Conference so, with nothing much to write about, I headed for Pleasure Beach and Sir Hiram's Captive Flying Machine. What gave me the idea was the memory of my father talking about the time he, as a small boy, had made the ride with *his* father (back in the days when a tolling bell indicated it was the ladies' turn to bathe; any man caught lingering on the beach risked being fined a bottle of wine).

I was unprepared for the size of the thing. The open cars, each accommodating eight passengers on four hard seats, were modest enough, but the structure itself had a looming industrial immensity, like an oil rig blown ashore in a storm. The first thing I noted was the absence of safety belts; there was nothing to hang on to but the sharp metal edges of the car itself. Two elderly Welsh women, possibly sisters, had done it before. One said, 'Nothing to worry about, love, centrifugal forces will keep you pinned down.' At first the car scraped along the platform then, steadily gathering speed, was thrust outwards and upwards by the whirling slingshot principle. Buffeted by the wind, going ever higher and faster, I found myself recalling a flight I'd once made in a Royal Navy Phantom jet. Then, when the main business of the day – a ninety-minute mock dogfight over the North Sea – had been concluded, the pilot sank to 200 feet and, idling along at two hundred and fifty knots, remarked that a Phantom, lacking guns, was vulnerable when all its missiles had been fired. 'In that situation,' he had said, 'there is only one thing we can do.'

'And what's that?' I asked.

'We go up.'

There was a roar as he ignited the afterburners, and a bang in the back like an elephant's kick. The noise, vibration and power were almost indescribable; rammed into my seat it seemed that, if we did not slow down soon, we were going to burst clean out of the atmosphere. The altimeter was spinning like a top when, eighteen seconds later, we levelled off at 25,000 feet, leaving me with the kind of high that, most oddly, I was experiencing again aboard the Captive Flying Machine – not quite the same intensity, obviously, but for something devised back in the Edwardian era, simply amazing – and an unexpectedly moving, yet exhilarating way (I was grinning inanely) of commemorating my father.

It remains, to this day, my best memory of Blackpool.

The town has numerous veteran rides – the Helter Skelter Lighthouse, the Hall of Nonsense (a mirror maze), the Whip, the Zipper Dipper and the Tumble Bug are among those that have been enjoyed for generations – but Maxim's Flying Machine remains the most venerable of all. And, though he built it for a fairground, his declared aim was 'To make the British public airminded.' Ours were not the original cars. Back then people sat in vehicles that resembled small aeroplanes – he'd even wanted to fit them with electrically driven propellers – in which the illusion of flight was so strong that, years before anyone in Britain had actually got airborne, thousands of exhilarated holidaymakers reckoned they knew exactly how it felt.

People admired Maxim. With his shock of white hair, his legendary gun and fabulous steam-powered flying machine, he was rarely out of the public eye. And the personalities he knew! The Royal Family, members of the Cabinet, famous writers, leading men of science – they all, eagerly, danced attendance upon him. Few, though, would have heard of his home town: Sangerville, tucked away in an obscure corner of Maine, or known that in 1840, the year of his birth, the greatest problem facing its citizens were the black bears that roamed at will through Piscataquis County.

Hiram's father, Isaac Weston Maxim, had cleared a few acres of forest on the outskirts of Sangerville where some of the biggest – up to five hundred pounds – and greediest tended to congregate. There he built a house where Hiram's mother, Harriet Boston Maxim, soon learned *not* to fry bacon for breakfast (the smell drives bears wild, they'll beat down the stoutest kitchen door). Hiram told his own son, Percy, that he'd been 'a poor little bare-headed, bare-footed boy . . . running wild but very expert at catching fish.' It's said of Ernest Hemingway that if he encountered a bear while walking in the woods he simply whacked it on the nose with a rolled-up newspaper. It seems young Hiram, fast acquiring a reputation for absolute fearlessness, dispatched them with equal nonchalance, a smack on the snout or clip on the ear being enough to send them shambling off through the trees.

At fourteen he began working for an East Corinth carriage-maker named Daniel Sweat (a hard taskmaster and butt of many family jokes) then, when Isaac gave up farming to manage a grist mill, took a job with his father. There is a picture of him at about this time showing a clean-shaven young man with thick black hair, a high forehead, large, intense eyes and a prominent nose. He wasn't handsome, but the face is interesting enough to make one look twice, and his full, rather sensuous lips offer just a hint of the scandal that would banish him to Europe. It was Isaac's exasperation with rodents that inspired Hiram's first invention – that ingenious mouse-trap which, after each kill, would automatically reset itself. (Years later, seeing it in a hardware store and learning it had become a mass-manufactured bestseller – widely regarded as the best trap ever devised – he threw up his hands in dismay; he had completely forgotten to patent the thing.)

He dreamed of travelling, of making a great journey, yet got no further than Fitchburg, Mass., where his uncle, Levi Stevens, owned a business building gas machines. These, in pre-electricity days, provided lighting, and here Hiram began showing signs of a ferocious engineering talent which, when he moved to New York, flowered into genius. Maxim & Welch, the firm he co-founded there aged only thirty-three, quickly became one of the nation's leading creators of gas and steam engines – the former, through no fault of his, often proving highly volatile. Indeed, after one exploded and set a neighbourhood factory alight, Hiram invented the world's first automatic sprinkler system, a brilliantly clever device that summoned the fire brigade even as it put out the fire. Yet he could find no backers, and when, in time, others built and successfully marketed his idea, he realized that, once again, he had neglected to take out a patent.

It would never happen again. In 1878 he filed his patent for 'Improvement in electric lamps' (the first of two hundred and seventy-one patents he would publish) then helped set up the United States Electric Lighting Company. Appointed its chief engineer, he designed generators and created arc lamps for public spaces (Hiram lit up the New York Post Office) even as he and his hated rival, Thomas Edison, raced to devise a simple, reliable incandescent lamp that would illuminate homes all over America. Both knew that whoever found the answer – the light bulb! – would win one of the great prizes of the nineteenth century, along with an assured place in history. Edison, late in 1879, got there first. But then Hiram, hot on his heels, came up with a better bulb containing a superior filament. Edison, ruthlessly juggling patents, filing lawsuits and exploiting the courts, was granted the freedom to manufacture his own version of Maxim's finer filament. Hiram never forgave him.

Meanwhile his personal life had begun attracting attention. In 1867 he had married his first wife, Jane Budden, in Boston, and fathered three children. In 1878, in New York, he married Helen Leighton. In 1880, also in New York, he wed Sarah Haynes. Then Jane divorced him on the grounds that, when still married to her, he had married Helen, while Sarah divorced him after realizing she had been shared with both Helen and Jane – that, at one point, scarcely credibly, he had been married to all three simultaneously. As the papers began getting wind of the story he learned that one of his inventions, an electrical pressure-regulating system, had won a Légion d'Honneur at the 1881 Paris Exposition. Invited over to receive it, he quietly boarded the SS *Germanic* and fled the country. But Sarah followed and, in 1890, in England (at South Norwood, near Croydon), married him all over again. Hiram, resisting any further adventures, remained with her for the rest of his days.

In Paris an unnamed American advised him that anyone wanting to make a huge fortune should 'invent a killing machine . . . that will let these Europeans cut each other's throats more easily.' That appears to have set him thinking. He came up with an idea for a gun which, if each discharging cartridge could somehow provide the power to load the next, would fire its rounds automatically. At the same time, persuaded to stay abroad by the drubbing Three Wives Hiram, or the Electric Trigamist, was likely to receive in the US press, he decided to make his home in Britain. Here, at least, they spoke his language.

No. 57D Hatton Garden is a nondescript brown-brick building standing on the corner with Clerkenwell Road. It bears a blue plaque which says,

Greater London Council
Sir HIRAM MAXIM
1840–1916
INVENTOR AND ENGINEER
designed and manufactured
THE MAXIM GUN
in a workshop on
these premises.

Today a sign on the fashionable Hatton Garden side identified the place as Kovaks House. A heavy smoked-glass door led to a tiny lobby in which I could just make out a vase of lilies and a girl seated at a desk. The door, when I tried it, was locked, but a speakerphone buzzed and a female voice said, 'Yes?' Peering through the glass I said, 'I'd like to speak to somebody about Hiram Maxim, please.'

'Who?'

'The inventor. He used to work here, at number 57D.'

'Never heard of him.'

'But he's on your blue plaque.'

'What blue plaque?'

I pointed upwards and to my left. 'This one. It's on your outside wall.'

'I don't know anything about *blue plaques*.'

'Well, perhaps there's someone in your office who does.'

'No one's available.' She was growing testy. So was I. Then I dimly perceived two words written large on the wall behind her: 'INTERNATIONAL BULLION'.

'Or someone who could sell me a sovereign.'

The speakerphone clicked. She had switched me off.

Hatton Garden might look narrow and dingy up here, but at the Holborn Circus end – where at least there were broad pavements and a few trees – stood a NatWest branch that had to be one of the busiest in the country. Yet the old Colonial Buildings nearby, raised during the nineteenth-century South African diamond rush, now, ironically, sported To Let signs. (Modern Hatton Garden has its origins in the world's largest manmade hole – I've peered, awed, over its edge – dug 1,097 metres deep in Kimberley by thousands of frantic men with spades; the two and a half *tonnes* of high-grade stones they produced, sold on by Cecil Rhodes and Harry Oppenheimer, caused jewellers to begin congregating here in the 1870s.) Today more than three hundred businesses are located in and around the thoroughfare.

I strolled past shops with names like Krystle Diamond Rings, the Imperial Pearl Company, E. Katz & Co. Ltd and Amber Hall Jewellery, in the window of Hatton Garden Pawnbrokers studied a lovely antique silver necklace displayed alongside some exquisite filigreed tortoiseshell combs and cigarette cases. Turning down Greville Street, popping into the Jewel Lounge, Cool Diamonds, Go for Gold, the Diamond Palace and Ace of Diamonds, I learned the top of a round brilliant-cut diamond is known as the crown, its middle the girdle and its bottom the pavilion, that flawless means not a single inclusion or blemish visible under ten times magnification, and that the best colour is very, very white. In one store a quietly spoken Asian woman told me about conflict diamonds. 'We call them the crack cocaine of the jewellery trade. The Kimberley Process, signed in 2000, was meant to put a stop to it, and if you're buying, you should always ask the jeweller if he subscribes.'

But places like Angola and Sierra Leone, presumably, didn't produce rubbish stones?

'On the contrary, some of their stuff is fabulous. But that's not the point.'

'And does any still end up in Hatton Garden?'

She shrugged. 'Of course.'

I asked several jewellers if they'd ever heard of Hiram Maxim ('He's on a blue plaque at Number 57') but none had.

All this land, as any London guidebook will tell you, had once belonged to the Bishop of Ely, who had a palace here. Then, in 1576, Elizabeth I gave one of her favourite courtiers, a handsome young man named Christopher Hatton, a house and grounds in which he planted a famous garden. In return he paid her an annual rent of

£10, along with ten loads of hay and, on Midsummer's Day, a single rose. In 1607 Hatton became Lord Chancellor of England; soon afterwards his garden was turned into streets and dwellings.

One of its more obscure corners, Bleeding Heart Yard, is a small enclosed cobbled space containing the Bleeding Heart Restaurant (with four hundred different wines in its cellar) and Topical Time Ltd, Watch and Clock Wholesalers. A man of about my age, wearing a beautifully cut coat with an Astrakhan collar, was heading for its entrance. He gave me a friendly nod and, on the spur of the moment, I said, 'Pardon me, but you wouldn't happen to know how this yard got its name, would you?'

'Ah!' he said. 'It's one of those stories you frighten the children with.' His voice, a silky, rumbling bass, had probably been seasoned by decades of hand-rolled Havanas and premium malts. 'One night in 1626 Lady Elizabeth Hatton, granddaughter of the sainted Christopher, was spotted dancing with a mysterious stranger. Next morning her body was found in this very yard, ripped to pieces, her chest slashed open with such violence her heart was left exposed and bleeding – but still beating.'

'Good God,' I said, and he looked gratified. Yet he too was stumped when I enquired about Hiram Maxim. Indeed, my poll yielded only a single positive response. Having corralled an infant who, lifted down from a big Mercedes, had made a wild, sudden dash for the road, I chatted briefly to its father – one of a number of curly-bearded, black-hatted Hassidic Jews moving around the Garden in a distinctly proprietorial way. 'Maxim?' he said. 'Of course. He invented the sub-machine gun right here, at number 57D. There's even a blue plaque.'

At 57D, in 1883, Maxim wrote, 'I began experimenting on the automatic gun, and, for it, I obtained my first English patent.' (The first English patent ever granted for such a weapon went to James Puckle who, in 1818, produced a gun which shot round bullets at Christians and square ones at Turks.) What made the Maxim so different from the multi-barrelled Gatlings and Hotchkisses was a lone barrel firing six solid-drawn brass cartridges in just half a second. It caused a sensation. The Duke of Cambridge, Commander-in-Chief of the British Army, called in for a demonstration, as did the Prince of Wales. 'This visit,' wrote Hiram, 'was followed by those of many other royalties, dukes and lords.'

He was a man of immense physical strength. There is a photograph of him holding, outstretched in one hand, the prototype of his famous weapon; he looks as nonchalant as if offering someone a rose, yet the Maxim gun was notoriously heavy, requiring from most a determined two-handed lift. His own son, Percy, recalled him throwing a burglar bodily over a fence, also leaping aboard a moving train to grab a man who owed him money. 'My father,' he wrote, 'was extraordinarily powerful . . . and as quick as a cat.'

If artifice was necessary to make a sale, he'd use it, for example, to impress Emperor Franz Joseph of Austria he spontaneously machine-gunned the initials FJ into his palace walls. Much of his ordnance went abroad. (Germany, during World War One, used a home-made version of the Maxim, while the Allies adopted its direct descendant, the Vickers; on his death he expressed no remorse at his contribution to the slaughter.) Some nations, though, chose not to buy at all. A pigtailed envoy from the Middle Kingdom's Dragon Throne watched a demonstration – held in the Cotswold garden of Hiram's City banker – but found it chewed up bullets at a rate his exchequer couldn't afford. Regretfully he declared, 'This gun fires altogether too fast for China.'

Maxim, who took British citizenship in 1899 and was knighted by Queen Victoria, returned to America infrequently. In 1914 Sangerville threw a big party for its celebrated expatriate at the Town Hall. Whether he enjoyed it is not known. All we can be sure of is that he never set foot in his birthplace again.

His interest in aviation seems to have stemmed from the moment when, aged sixteen, he found a drawing his father had made of a

rudimentary helicopter. Isaac's fascination with flight proved conta-
gious, and from then on it became a topic they frequently – often
heatedly – discussed. In 1872 Hiram drew a helicopter of his own,
but soon abandoned it in favour of a fixed-wing machine. In 1887
someone asked him if such ideas were not mere pie in the sky. 'The
domestic goose,' he said, 'is able to fly, and why should not man be
able to do as well?' He estimated it would take five years and cost
£100,000 and, backed by his armaments fortune, decided he might
as well be that man.

Needing a spacious, fashionable property befitting someone of
his station, he chose Baldwyns Park in Bexley, Kent. Bexley proved
perfect for his requirements. Thoroughly pastoral, with fields and
hedgerows, country fairs and a country church topped by a shingled
spire, it lay only fifteen miles from the City of London. He moved
there happily, and today deserves mention alongside such celebrated
residents as William Morris and Admiral Sir Clowdisley Shovell.
(The latter, having terrorized the French into scuttling their fleet at
Toulon, was wrecked off the Scilly Isles in 1707; Shovell, immensely
fat, and wearing an enormous emerald ring, came bobbing ashore
like an empty cask, but then, as he lay semi-comatose on the beach,
a local woman smothered him – probably by sitting on his face –
before amputating a finger and stealing the ring.)

At Baldwyns Park, Maxim built a giant hangar and a wooden
wind tunnel in which twin steam-driven propellers provided a steady
air flow. Having rigged a hand-rolled brass wing in the tunnel, he set
the propellers to blow at precisely 40 mph and started measuring
lift. Then, to better understand it, he watched birds (their use of air
currents, particularly thermals, interested Hiram – who, according
to James E. Hamilton's excellent book *The Chronic Inventor*, once saw
ten thousand barrels of alcohol in a New York bonded warehouse
create a super thermal when they went up in flames; the heated air
rose so fast that trash sucked from the surrounding city rose with it,
while cool air rushing to fill the resulting pressure void did so in the
form of gale-force winds).

Maxim's machine, despite its great size, looked virtually weightless.
Fashionable London society turned up to admire it, while Kipling,
H. G. Wells and the Prince of Wales were among those who
begged for rides.

His flying machine, he decided, would be propelled by steam and
launched from an eighteen-hundred-foot-stretch of specially built
railway track with a nine-foot gauge. Boasting that nothing would be
used in its construction but 'a carpenter's two-foot rule and a grocer's
scales', he went to work, produced first a giant open central section
containing a bridge or command deck with, high overhead, a graceful forty-foot hexagonal sail strung laterally, like an awning. After he
fitted the wings his machine had a span of a hundred and four feet
(a modern DC9 Series 30 passenger jet has a wingspan of ninety-
three feet four inches).

The boiler, carried in the nose, was a marvel of precision
engineering. Though only eight feet long, six feet high and four feet
wide, it boasted eight hundred square feet of heating surface, and a
naphtha burner with 7,560 jets each producing a twenty-inch-long

flame. Steam was piped to two featherweight engines that turned two eighteen-foot twin-bladed propellers – each painstakingly sculpted from American white pine then covered with Irish linen and varnished till they shone. On the bridge a wheel controlled the silk-covered rudders, also the front and rear elevators, while a throttle lever led to the engines, both positioned aft. Propellers at the front (the horse pulling the cart) did not interest him. 'What is true of ships,' he wrote, 'is true of flying machines.' He talked of the backwash frontal propellers would cause, a speeding-up of the airflow merely serving to increase wind resistance.

The wheels weighed just over a ton.

Yet his finished machine, though awesomely big and beautiful, looked almost weightless. Virtually every magazine and paper in London ran respectful pieces, while his good friends H. G. Wells and Rudyard Kipling published their tributes too (Wells even based a well-known short story, 'The Argonauts of the Air', on Hiram). Griffith Brewer of the *Morning Post* summarized the general mood when, on 5 November 1892, he wrote, 'There can be no doubt that Mr Maxim's machine is the nearest approach to a practical flying machine yet made, and even if it does not succeed in flying he will have advanced the science of aeronautics to a considerable extent.'

He made many test runs using only the huge centre section. Once, as he puffed along at 35 mph, a small breeze lifted the machine clear off the metals and dumped it in the field alongside, necessitating repairs costing £1,000. Safety rails of Georgia pine were set, like a low fence, thirteen feet outside the track, and outriggers fitted to his wheels so that if they lifted again the pine rails would catch the outriggers and force the machine back to earth. This measure meant, of course, that if Hiram ever did fly he could only rise an infinitesimal distance, yet large crowds turned up to watch the trials, while dignitaries such as the Prince of Wales, H. G. Wells and Admiral of the Fleet Sir John Edmund Commerell hastened to bag rides.

On none of these outings, however, did Hiram attempt to get airborne. First he extended the track to almost half a mile, then, deciding even more distance was needed, asked that a small wood be cut down. Balwyns Park, however, was a rented estate, and his landlords advised that the cost would be £200 per tree. He withdrew his request.

On Tuesday 31 July 1894 the machine, weighing over three and a half tons (including 600lb of water), was finally judged ready. Hiram and his three crewmen climbed aboard and took their positions on deck. He opened the throttles a small way and, accompanied by a great noise of steam (but surprisingly little vibration), those massive propellers began to turn. When, trundling along at a good rate, there was no sign of lift he opened them further then, judging his moment, further still. What happened next took them by surprise. 'I ran up the steam pressure,' he wrote, 'until the screw thrust exceeded 2,000 lbs and then when the engines . . . were developing this tremendous amount of power, the machine . . . darted forward with a suddenness for which the crew of four men were not quite prepared.' (A couple were thrown to the deck.)

At 42 mph, lifting sharply, they broke the restraining safety rail 'so that the machine was liberated and floated in the air, giving those on board the sensation of being in a boat.' When a length of snapped rail snared a propeller 'I shut off the steam, and the machine stopped and settled to the ground.' It came down, in fact, so hard that a crewman, Thomas Jackson, bounced out and landed on his head; miraculously, he sustained only severe cranial bruising. They had flown for a distance of six hundred feet and, until Orville Wright came along nine years and a hundred and thirty-nine days later, held the world records for duration (eight seconds) and altitude (two feet).

Then, even as he repaired his machine and considered further trials, Baldwyns Park was sold to the London County Council – who, in turn, planned to turn it into the Bexley Lunatic Asylum; Hiram remarked dryly, 'It appears that I had prepared the ground, so that all that was necessary was to erect the buildings.'

Yet it wasn't just at Baldwyns that the builders went to work. Property refugees driven from London by London prices would soon

Periodically, demonstrating his support for local good causes,
Maxim held open days at Baldwyns Park. Crowds flocked to see his
flying machine – and guns.

begin urbanizing Bexley village and its rural surrounds. Maxim's
'prestigious' estate is today filled with neat, well-maintained executive homes, while the spot where he probably had his workshop, on
the corner of Baldwyn's Park (now a road) and Old Bexley Lane, is
occupied by Top Notch Dry Cleaners. When, early in the twenty-
first century, Bexley built a new mental hospital, Maxim's early
nineteenth-century manor house – Ionic columns flank its entrance
– was converted into 'exclusive apartments'. While the man who'd
helped invent the light bulb might have puzzled over estate agent's
details such as 'low-voltage halogen uplighters', certain things would
have been familiar. The original block stone floors had been
retained, while the only addition to the cellars – where I pictured
him, oil lamp in hand, selecting a nice burgundy for dinner – was
electricity.

On a fine spring day I strolled past the building which, from the
outside, seemed largely unchanged – though I noted five big dormer
windows that hadn't featured in period photos. The gardens were

well kept, I smelled fresh-cut grass, and, had it not been for the proximity of the Bexley Park Sports & Social Club, it wouldn't have been
a bad place to live. The club, a sprawling, flat-roofed, single-
storeyed, institutional-looking structure of quite remarkable ugliness,
was plonked so close to Baldwyns' blue-painted double front door
I guessed it had once been part of the asylum. A tall man walking
a yapping little terrier confirmed this. 'Dunno which part, though.
It's all been recycled now; down the road they've even turned the
chapel into a fitness centre.' At the club's rear a patio with tables
and chairs looked out across a playing field, and it suddenly dawned
on me that this would have been the ideal spot from which to sit and
watch Maxim getting airborne. Pictures of the time had the railway
track running just beyond the present playing field (and past great
stands of oaks). One in particular, showing the house, the aeroplane
and dense, milling crowds, was taken on an afternoon dedicated to
raising money for local causes. A contemporary poster gives us a fair
idea of why all these people had come. It was headed:

*Bexley Cottage Hospital*
Exhibition of Mr. Maxim's
FLYING MACHINE & GUNS

He always looked back on his Baldwyns Park adventure with pride
and, years later, would brag, 'The late Mr. Cody, who was one of the
most skilful men on kites and flying machines that we have ever had
in England, just before his unfortunate accident, said: "Sir Hiram,
your machine was all right – the proportions were all right; it was
not too large, and I'm going to make another one just like it, with no
change except in the motor-power." '

He and Cody, though both obsessive, were otherwise direct
opposites in character – top-hatted Hiram, rigid, stiffly starched
and ultra-conservative, was notorious for his boastfulness and self-
aggrandisement ('I doubt the goodness of his purpose,' wrote Wilbur
Wright, 'and dislike his personality'), while Cody was the long-haired
showman, easy-going, self-deprecating and funny and, despite his

prodigious physical strength, possessed of a gentle, companionable nature. Yet their relationship, based on common interests and a shared nationality, was close. In 1910 they shared also some serious concerns about aeroplanes being used as instruments of war. Hiram, who thirty years before the Blitz anticipated enemy bombers crossing the Channel to drop tons of nitroglycerine on London, lobbied Winston Churchill, then President of the Board of Trade; Churchill, in turn, was concerned enough to organize a meeting between Hiram and the War Office's Aerial Navigation Sub-Committee. His message – being spread with equal vigour by Cody – was that aeroplanes must be fought with aeroplanes: Britain would need fighters to shoot the bombers down.

Meanwhile he was unaware of the effect he'd had on a school-boy named Geoffrey de Havilland who, years later, would write, 'I was twelve when he was carrying out the trials with his extraordinary machine laid on rails in Baldwyn's Park in Kent, and eagerly read about them at the time. My friend and co-director, C. C. Walker [Charles Walker, de Havilland's closest associate] did better than this, though, and actually went on board for a test. He can confirm that the upper guard rails to prevent the machine from taking off were distorted, so proving that it had lifted.'

Hiram, having written a book, *Artificial and Natural Flight*, that few people ever read, continued working on projects ranging from coffee-roasting machines to bulletproof waistcoats. Towards the end of his life, weighing seventeen stone and suffering from severe bronchitis, he devised a glass inhaler which proved so effective that, when a London firm manufactured it, 'hundreds of thousands' were sold worldwide. Hiram died in 1916, at a moment when he was endeavouring to turn kerosene into petrol, and Sarah, Lady Maxim, having married him twenty-six years earlier in the South Norwood registry office, chose to bury him in the West Norwood cemetery. And there to this day he lies, largely forgotten.

## Chapter Six

# Sky Fever

Geoffrey de Havilland

On 9 February 1911, *The Times* published a letter from the rector of Crux Easton, an obscure little parish located some six hundred and fifty feet up on the north Hampshire Downs. 'Now that aviation is making remarkable progress in many countries,' wrote Charles de Havilland, 'I beg to make a solemn suggestion. There is a clause in the Litany used in the Church of England which is as follows: "That it may please Thee to preserve all that travel by land or water."'

Hadn't the time now come, he enquired, to add the words '*or by air?*'

'May one further hope that the Archbishop of Canterbury, the Pope and the Head of the Greek Church may simultaneously give instructions for that addition to this beautiful form of prayer?

'I'm afraid I must plead personal interest as being the origin of my suggestion, which has arisen owing to my son having taken up aviation and being now at the Balloon Factory, Farnborough.'

A year earlier, on a bright spring morning in 1910, Revd de Havilland had been present on Beacon Hill (just three miles from his Crux Easton rectory) when the boy made his first flight – though, cowering behind a shed, unable even to peep through his fingers, he failed to witness the actual event. First Geoffrey, aged twenty-seven, climbed into a garden chair roped to the fuselage of his skeletal little aeroplane and signalled to his mechanic, Frank Hearle, to swing the propeller. Next the barking, smoking four-cylinder engine, built to his specifications by the Iris Motor Company of Willesden, took him bumping up to the prehistoric hill fort on the summit. 'It was a good place to go flying,' Geoffrey later recalled, 'and I felt no premonition of danger.' Then he turned, opened the throttle and went plunging down until, judging his moment, he 'pulled back hard on the stick'. As the machine rose the wings snapped off. It hit the ground with a loud, oddly musical report – snapping piano wire audible above the percussive whump of splintering wood. For a moment Geoffrey sat too stunned to move. Then, noting that Hearle and his brother Hereward were running towards him – while his father, impelled from his hiding place by the noise of the crash, puffed along behind – he raised a hand to indicate he was unhurt. But the propeller, still windmilling behind, gave it a ferocious whack and, as they arrived, he stumbled from the wreckage bleeding copiously.

The Revd de Havilland, 'speechless with shock', was eventually persuaded to go home. Geoffrey knew instinctively his wings had failed because certain spars were cut from American white pine, featherweight but frangible; hickory or ash would, almost certainly, have kept him up longer. He and Hearle raced back to their work-shop in Bothwell Street, Fulham, and began work on his second aeroplane. It was this elegant, supremely airworthy machine, launched late that year from Beacon Hill, that won him an attachment to the Balloon Factory, established his reputation and got him talked about – even at Crux Easton's village school. There, according to the Register, on 1 December 1910, 'the Rector came this morning to tell the children about the success of his son "with regard to his aeroplane" in which they have taken a great interest. To celebrate the event a half holiday was granted.' (Yet they got none on 5 February 1912 when, as playground puddles turned to ice and 'the ink was frozen for the greater part of the day', they took a great interest in just trying to keep warm.)

Geoffrey de Havilland, born in 1882 at Terriers Wycombe, Buckinghamshire, was barely out of his pram when his father was appointed to a parish in Nuneaton. Towards the end of the family's fifteen-year sojourn there – he had two brothers and two sisters – his mother, Alice, suffered a nervous breakdown. Raised in a tranquil, willowy corner of rural Oxfordshire, she had never felt comfortable in this cold Midlands town with its unwelcoming, pinch-faced, misogynistic citizenry. (The only way Mary Ann Evans, Nuneaton's best-known resident, could make her way was by pretending to be a man, George Eliot.) Also, the family kept lurching through a series of financial crises caused primarily by Charles's curiously exalted sense of his own worth. He believed that tradesmen and shopkeepers could either wait for payment or, better still, write off entirely what was owed – not because he was a poor churchman (which might gain them merit) – but because he was descended from Lord De Havyland, one of William the Conqueror's Norman knights. It was his children, sent off to the shops without a penny piece between them, who bore the brunt of the abuse and returned home in tears. The personality of her husband was, in fact, the key to her troubles.

Born with one leg shorter than the other and forced to wear a large orthopaedic boot, he was known for his uncertain temper (once, when a shoulder of mutton proved difficult to carve, he hurled it into the fireplace) and 'terrible outbursts of shouting'. Yet this edgy, difficult man found a measure of peace in his garden where, assisted by Clench, a toothless Crimea veteran who owned a muzzle-loading gun, he raised Muscovy ducks and William pear trees. The eggs and fruit were always brought to his study and arranged in neat rows, none could be touched until he judged them ready for eating – yet he never did. Neurotically incapable of allowing his children these small pleasures, he elected to let the food rot instead (it was their mother who, covertly, cleaned up the mess). Much of his adult life was spent, apparently, writing a book on the locations of certain Biblical places which, he claimed, the Bible had got wrong; they all grew accustomed to hearing how, on publication, the entire history of the Middle East would need to be reappraised. Yet, on his death, Geoffrey found he had scribbled no more than a page or two.

The family member for whom Geoffrey felt most profoundly ('the most important figure in my life') was his maternal grandfather, Jason Saunders. A wealthy businessman – for many years he ran the Oxford Tramways Company – and active member of the Liberal Party, Saunders was also the ministering angel who kept bailing them out when money was short. As his fortune grew, so did a passion for local politics so profound they even carved his likeness – Alderman Saunders with the broken nose and bicorn walrus moustache – into a corridor of the Oxford Town Hall. There he remains to this day, glaring down like an old fairground prizefighter, yet he was a famously generous and principled man who, after being elected Sheriff of Oxford, went on to claim Oxford's Mayoral chain.

Geoffrey adored him and, in his autobiography, *Sky Fever*, recalled Medley Manor, Saunders' large Thames-side farm, with a kind of elegiac reverence: there was Harris, the carpenter, 'knee-deep in sweet-scented wood shavings', and Druid, the white-bearded wheelwright, and Cox, the blacksmith (with his one-eyed assistant). Years on he could still hear 'the musical notes of hammer on anvil, the hum of the chaff-cutter, and recapture the warm scents of the garden, the bed of violets, the earthy smell from the potting shed and the scent of peaches and nectarines from the hot walls which seemed to reflect the glow of perpetual sunshine. It was a heavenly place.'

The farm, located in Binsey Lane off the Botley Road – along which undergraduates once rode to hounds – today produces a selection of pick-your-own fruit and vegetables such as strawberries, rhubarb, carrots, beetroot and, reportedly, the best asparagus in Oxford. Hearing it was owned by a Mr Charlie Gee I called and learned there were actually two of them, a dynastic father-and-son arrangement. 'It's my dad you want,' said Charlie Gee Jnr, and an appointment was made. I imagined Medley Manor's entrance would feature tall pillars and ornate wrought iron but, arriving on a fine spring morning (hawthorn in bloom, songbirds in every tree), found a simple farm gate propped open – along with a scrawled notice regarding asparagus availability. The house, a handsome, spacious, squared-off structure three storeys high, lay at the end of a long, straight drive. Charlie Gee Snr turned out to be a fit-looking old countryman with sparse grey hair and wire-rimmed glasses through which he examined me in a calm and measured way. When I explained my interest in de Havilland he smiled and said he'd actually met him. Twice.

'It was back in 1960,' he said. 'Somehow he got hold of my name and wrote me a letter. It was about Jason Saunders, and how he'd spent his summers here as a kid, and asking if he could have a look around. One morning a whacking great Rolls-Royce drove up and out he stepped, the man himself. Joan, his second wife, was with him.'

'What was he like?' I asked.

He considered a moment. 'A tip-top gentleman. Unassuming, quite shy, I thought. Only spoke if he had something to say – though he had a good deal to say here. There's a landing on the stairs with a built-in cupboard, and finding that cupboard really pleased him; it was where his grandmother had kept her gooseberry jam! And when he saw the weathercock on the stable roof he said it had been designed by his mother; its outline traced on a piece of paper which Cox, the blacksmith, then cut out in iron. He showed me the spot where they used to keep the steam engine, and the ditches where he caught pike, and the lily pool where there was roach and perch. And he asked about an elm which, as a boy, he thought was the biggest tree in England; I had to tell him we'd cut it down. Then there was a sycamore his mother had climbed as a girl; it was still standing, thank heavens. He photographed it and, in fact, it was the main reason for his second visit. He wanted more pictures of that tree.'

He showed me several big shiny old-fashioned black-and-white enlargements of pictures sent to him by Geoffrey. 'He said it was while wandering around the fields here that he got interested in natural history; actually, I had the feeling that, if he hadn't become even more interested in engines, he might have dedicated his life to it.'

'His dad wanted him to go into the Church,' I said. 'But that was *never* an option.'

Now Charlie produced two flimsy pages, closely typed. 'He also gave me this. It's a copy of a letter sent by Jason Saunders to Christ Church, the Oxford college. Saunders actually rented this place from Christ Church. They still own a lot of land around here.'

'Do they still own Medley Manor?' I asked.

He smiled faintly. 'I bought it off them in 1957.'

The Saunders letter, basically a tenant-to-landlord list of complaints – the house, for example, being 'extremely cold and draughty, not to say damp' – described a Medley very different to the enchanted place Geoffrey had known. Yet, according to his book, some of that enchantment still lingered when he returned in 1960 and, though Charlie is not named, Geoffrey was full of praise for the way Medley had been maintained. 'Quite recently,' he wrote, 'Joan and I went to Oxford to look at Medley again. We dreaded the shock we felt sure was in store for us.' But, standing there with a very young Charlie Gee, he saw there'd been nothing to fear, and reflected, 'As I looked around me I knew that it was here that I had spent many of the happiest days of my life.'

He told Charlie his life had, in fact, come full circle – here he was, back at Medley, revisiting the best parts of his childhood – and Charlie told me how once, on holiday in Caernarvon, he'd come across an old plane that was offering joyrides then, learning it was a *de Havilland* Rapide, had gone up at once. 'It was a real adventure!' he beamed.

I left this engaging man and headed on to the hamlet of Binsey, half a mile away. It possessed a seventeenth-century pub named the Perch, a green where householders grazed their geese, a broad trout-bearing stretch of the Thames (on its far bank white horses wandered over the largest meadow in England) and a tiny thirteenth-century church with a miraculous spring that had attracted pilgrims through-out the Middle Ages. During the Victorian era it came to the attention of Lewis Carroll, a Christ Church mathematics don, who'd called it the Treacle Well, and made it part of a wildly surreal tale told to Alice, by a dormouse, at the Mad Hatter's Tea Party. It lies, a muddy puddle, just yards from the grave of Mary Prickett, Alice Liddell's governess, later immortalized by Carroll as the loopy old Red Queen in *Through the Looking Glass*.

I could only imagine, standing in that quiet churchyard, how badly Jason Saunders had wanted Charles to be appointed curate here. Close to Medley – a mere ten minutes by pony and trap – and set in the lush, privileged, familiar corner of England where his daughter had grown up, it would have been the perfect antidote to her joyless Nuneaton years. The problem, I suspected, had been Revd T. J. Prout, the existing Binsey curate. His affection for the small, ancient church with its soggy little well has been documented; he would never have been persuaded to move.

In the event the nearest available living was about forty miles away. In 1896 Saunders purchased, for Charles, the advowson at Crux Easton, Hampshire, a hamlet consisting of a 'rectory, village school, church, farmhouse and less than a dozen scattered cottages'. At least here Alice had green fields and country air yet, Geoffrey noted sadly, 'I doubt if she was any happier.'

She might, secretly, have missed Nuneaton's bustle and vibrancy. Crux Easton was set along a quiet country road, its most interesting features a wind engine in Well Meadow (which pumped water and ground corn), and a copse containing the remains of a seventeenth-century grotto about which the poet Alexander Pope had written some very bad verse – 'This radiant pile' he called it.

The rectory had nine bedrooms (but one bathroom), along with a benign ghost in the cellar, enormous grounds and some interesting outbuildings. A century-old vine clambered over the house's south-facing wall so that in summer, if you wanted grapes – which were famously sweet – you just leaned out of a window and picked them. In time Geoffrey grew to love it almost as much as Medley, though, when he and his brother first arrived, it was still illuminated by oil lamps. Brilliant, charismatic Ivon knew about arcane subjects like electricity (while at Rugby he'd helped wire up the school) and Saunders had insisted his daughter must have this newfangled convenience too. So the brothers made it the first electrified house for miles around, starting with a small paraffin-powered engine emitting a light feebler than a glow-worm's, replacing it with a coal-fired boiler which needed stoking and triggered blackouts (Charles believed these were caused by electricity pouring wastefully from empty light sockets) then installing a big Crossley oil engine. This produced a radiance so intense it could probably be seen in Salisbury.

Geoffrey, for reasons never made clear, was sent not to Rugby but to St Edward's, Oxford,* a minor, if architecturally imposing, public school on the Woodstock Road. He hated the place with its 'hard and unapproachable Warden', but then in 1899, aged seventeen, moved to a parsonage near Gloucester where less academically gifted boys

---

* Geoffrey was not the only St Edward's boy to make his name in aviation. Group Captain Douglas Bader (who had both legs amputated at twenty-one yet still flew Spitfires against the Luftwaffe), and Squadron Leader Guy Gibson VC (leader of the dambuster raids that paralysed the Ruhr) were both Old Boys. Gibson died when his Mosquito, built by Geoffrey, ran out of fuel and crashed near Steenbergen, Holland, in 1944.

were prepared for university. Entry to Oxbridge, Charles thought, would guarantee him continuing membership of the ruling class; what he failed to understand was the classless nature of the world his son wished to enter. There it didn't matter what family you came from; all that mattered was how good you were – and later, among the flyers, how brave.

He had grown into a slim, dark-haired young man, with a long face, deep-set eyes, a strong jaw and large nose, his features combining attractively to give him an actor's looks and presence. Though his parents seem to have contributed little to his appearance – a Crux Easton parishioner recalled them as 'a red-faced, thickset man with a small frail wife' – his two female cousins were proof that the de Havilland bloodline could turn up some interesting surprises. Charles's half-brother Walter was a lawyer who had two startlingly beautiful daughters. Both moved to California and became movie stars.

Olivia de Havilland made her name in *Gone with the Wind* and Joan Fontaine in *Rebecca*; both were nominated, in 1941, for Academy Awards – sibling rivalry played out on a world stage; Joan won, though Olivia later went on to collect two of her own. During World War Two Geoffrey received food parcels from Hollywood's best grocers, in the Fifties his glamorous cousins came to visit. Delighted to find that both held pilot's licences, he gave them rides in his revolutionary new airliner, the jet-propelled Comet.

During his time at the Gloucester parsonage a local bicycle shop acquired two new 3½ hp Benzes and made them available for hire. Geoffrey and a fellow student saved up and booked one (the cost included a driver) to take them to Newbury. Neither had been in a car before, and when its leather transmission belts slipped on Birdlip Hill, both jumped out and pushed exultantly. At Newbury they were 'surrounded by a crowd of excited sightseers'. Geoffrey's 'great adventure' had an unexpected outcome. 'After that short drive I knew that my future life lay in the world of mechanical travel.'

Geoffrey with Olivia de Havilland on the flight deck of his
jet-propelled Comet airliner. Both Olivia and her sister, Joan Fontaine,
were qualified pilots.

Several months later, after intensive lobbying by his sons, Charles
bought a Panhard-Levasseur. Paid for, naturally, by Grandfather
Saunders, it had solid rubber tyres at the rear and, in front, pneumatic
ones which kept exploding. So grotesquely top heavy – passengers
sat in a big glass conservatory – that it tended to tumble over when
cornering, this Panhard was one of the most unstable automobiles
ever built. It was also one of the most dangerous. The engine was
started by lighting a small Bunsen burner beneath a system of metal
tubes and, when the tubes grew red hot, *pouring petrol into them*. Ivon,
incredulous, tore out the tubes and fitted electric ignition instead.
Shortly afterwards he joined the Oxford Electric Light Company as
a consulting engineer while Geoffrey, aged eighteen, began a two-
year course at the Crystal Palace Engineering School.

The 1851 Great Exhibition, an extravagant celebration of

Britain's wealth, power, invention and influence ('Foreigners also came, their bearded faces conjuring up all the horrors of Free Trade,' shuddered *The Times*), was held in a glittering, quarter-mile-long glass edifice erected in Hyde Park and dubbed 'the Crystal Palace' by *Punch*. It should have been demolished when the event ended, but the public felt such affection for it (in marked contrast, a hundred and forty-nine years later, to its feelings for the Millennium Dome) that the entire structure was moved across London to a wooded two-hundred-acre site on the summit of Sydenham Hill. In 1872 a School of Engineering was established at its southern end.

Geoffrey loved it. The Palace itself – to which every student received a free pass – had beautifully maintained gardens, a roller-skating rink, a library, a monkey house, a parrot house, even a Victoria Cross Gallery. There was live entertainment too: orchestral concerts, organ recitals, a variety theatre, regular choral festivals, prizefights, a famous flea circus, balloon ascents (he witnessed dozens) and, in the summer, firework displays staged by Brocks. The most popular of these was the Battle of Trafalgar, in which fiery warships exchanged fiery broadsides until the Franco-Spanish fleet, burning vigorously, vanished beneath the dark waters of the boating lake and, high overhead, a giant, fiery Union Jack flapped from *Victory*'s masthead.

But Geoffrey missed the most memorable display of all, put on by the Palace itself a full three decades after he'd gone. One evening in 1936 a small fire in a staff lavatory was allowed to spread, and within half an hour the flames could be seen right across London. Croydon Aerodrome enjoyed one of its busiest nights ever – the lights of circling aircraft, chartered by sightseers, creating an orbiting con-stellation in Sydenham's smoky red skies. At dawn, as firemen poked through the ruins, it became clear that only the School of Engineer-ing had survived.

It's still there. Now, though, it's where they keep those Palace memories alive – but only on Sundays and public holidays. When, one morning, I disembarked at the two-level station created for the two million Victorians who came here annually, no more than a few dozen left the train; most, carrying picnics, made for Crystal Palace Park. Meanwhile I trudged up Anerley Hill, the park's steep southern perimeter, to a small, square building of weathered greyish brick with a sign outside saying 'Museum'. Pausing a moment to catch my breath I noted that had it not been for a summery haze that made the light thick as honey, and the air almost sticky enough to stir with a spoon, the views would have gone on forever. Then I entered a modest-sized room boasting display cases, a counter heaped with books and a table containing glasses, wine, lemonade, peanuts and crisps. Two men and a woman, chatting animatedly, paid no attention to me. Had I gatecrashed a party? Finally breaking into their charmed circle I learned they were, in fact, the curators of the Crystal Palace Museum. 'We're volunteers,' the woman emphasized. 'That's why we don't open on weekdays. We've all got jobs.'

And those drinks on the table?

'It's our way of welcoming visitors. Mind you, we're a charitable trust, so we hope that anyone helping themselves will leave a small donation.'

I asked if anything was known about Geoffrey de Havilland. One of the men said, 'Well, we know he didn't like lectures – and this room was, in fact, the lecture theatre. He preferred the drawing office and workshop, now demolished, or better still, surveying, which took them out into the grounds. There they formed themselves into tribes and had pitched battles with ranging rods – using them like quarterstaffs, one lot tried to drive the other into the lake. Remember, it was a school of *practical* engineering. That's why he chose it.'

Trimly dressed in jacket and tie – unlike his colleagues in their baggy, unbuttoned weekend leisure wear– David Britton spoke with the intensity of the true enthusiast. He fetched a photocopied paper from the counter. 'This is a page from a 1905 edition of the *Crystal Palace Magazine*. You can see why the College appealed to a lad like Geoffrey.'

I read, 'Engineering is undoubtedly one of the prizes of the professional world, offering, as it does, plenty of scope for energetic men . . . [with] a taste for travel and adventure.' I learned that among the graduates who'd proved this rule were G. M. Hannan, 'Administrator of Martial Law in South Africa', D. C. Beringer, 'in charge of Cecil Rhodes "Cape to Cairo" Telegraph Party', and W. T. C. Beckett, 'district engineer Bengal and Nagpur Railway', while the student body included 'Colonials, Indians, Chinese, Japanese, West Africans . . . Egyptians, &c., most of whom left their countries without the slightest knowledge of engineering, and returned thoroughly skilled in its application, and *au fait* with its apparently inscrutable mysteries.'

David said, 'You're interested in the early aviators? Well, one of them, a wealthy young chap named Claude Grahame-White,* invented aircraft-carrier landings on two of our tennis courts, replacing the wire netting between with a restraining rope designed to catch his rear wheel. What would the length of two adjoining courts be? About eighty yards? Anyway, it worked. And today, even on those enormous American nuclear carriers, the planes still touch down using a principle created here at Crystal Palace.'

I was able to tell him that in 1905 Sam Cody, on leave from the Balloon Factory at Farnborough, built his first glider at Crystal Palace, then flew it in the presence of his friend and mentor, Sir Hiram Maxim. (Also that half a century later the foreword to Cody's biography, *Pioneer of the Air*, would be written by his friend and admirer, Sir Geoffrey de Havilland.)

He smiled. 'Care for a glass of wine?'

'It's a bit early for me.' As I left, he said, 'Next to the Engineering School, just yards from where we're standing now, was the School of Art, Music and Literature. It was lost in the fire, of course. Not many people know that Sir Arthur Sullivan was Professor of

* In 1910, while visiting Washington, Grahame-White landed his flimsy little Farman in the grounds of the White House and invited President William Howard Taft along for a ride. The President, who weighed twenty-one stone, politely declined.

Pianoforte and Ballad Singing there – fortunately he was away at the time.'

At the end of their courses the Crystal Palace engineers could, if they wished, undertake a special project. Since it offered an extra term in this magical place Geoffrey seized the chance, and for his project chose to assemble a 1½ hp motorcycle engine from drawings in the *English Mechanic* magazine. First, though, he went to see a shoemaker named Harris who owned a shop in Norwood and, as a sideline, built bicycles. Small, mild-mannered, perennially overworked and married to a woman who grew violent when drunk – more than once he turned up for work sporting a black eye – Harris often expressed concern at the way she abused his two children from a previous marriage. So when Geoffrey saw the eldest, a girl of eight dressed in rags, being 'beaten about the face and head', he 'thought it time to write to the N.S.P.C.C. After that things improved.'

(*He wrote a letter?* Why he failed to confront the woman and stop the beating remains a mystery – perhaps he didn't wish to distract Harris from completing the sturdy, high-quality bicycle that would be powered by his *English Mechanic* engine.)

In 1902 motorcycles were still a rarity in Britain. Crux Easton, when he declared it ready for his maiden ride, fairly bubbled with excitement. His parents, along with his sister Ione, brother Hereward and most of the village (including the school), were in attendance as, at the head of the lane that led down to the rectory, he took a deep breath and swung into the saddle. But when the bike gathered speed, its brakes failed. He elected to jump and, later, ruefully recalled the lane was surfaced with loose flints so wickedly jagged – his knees and hands got the worst of it – that he was laid up for several weeks.

He left Crystal Palace with regret – and no qualifications whatsoever.

Ivon was now designing cars at Daimler; Geoffrey, always influenced by his older brother, decided he must design cars too. He joined Wolseley at Adderley Park, Birmingham 'as a thirty-shillinga-week draughtsman'. The managing director, Herbert Austin, was an irascible, rather crude man, much given to shouting abuse at his staff, but also clever and supremely ambitious. He left to found the Austin Motor Company, created the phenomenally successful Austin Seven, became a multimillionaire (he paid himself a pound for every car sold) and peer of the realm. Geoffrey reported that the Wolseleys had horizontal engines driving clanking, convoluted systems of chains, which he judged to be 'reliable but noisy'.

But Geoffrey didn't like Birmingham – 'an overgrown Nuneaton' – or Wolseley's 'harsh régime' and, after just a year, resigned; other than having vague thoughts about being 'master of his own destiny' he hadn't the faintest idea what he would do next. Ivon, meanwhile, had become chief designer for the Iris Company. His first car, remarkably innovative and elegant, was one of the stars of the 1905 Motor Show, and as Geoffrey sat in Crux Easton considering his options, he was invited to Olympia to help man the Iris stand. But Ivon, laid up with a chill, was unable to be there, so at the end of the day Geoffrey called on him at his house in Acton.* Ivon, in excellent spirits, heard about the intense interest generated by his new saloon – Geoffrey had been bombarded with questions – and talked animatedly about the future.

Several nights later Geoffrey dreamed that he'd found Ivon lying motionless on the floor of his room. He awoke in a disturbed state, lit a candle and left it burning till morning, at breakfast, still shaken, told his mother. Shortly afterwards the postman brought her a telegram. She read it then, ashen-faced, gasped, 'Your dream!'

Ivon had died in the night. The chill may have become influenza, his condition perhaps exacerbated by overwork or 'emotional trouble'. Though Geoffrey, strangely, never gives a reason, let alone mentions

* Then situated on the Oxford road, Acton – 'oak town' – was a minor spa known for the soft water that bubbled from its springs. In the nineteenth century, foaming with soapsuds, it was the place where Londoners had their washing done – a hundred and seventy laundries operated in South Acton alone.

a coroner's verdict, his sense of loss was overwhelming, and for years he would be haunted by dreams – distant yet powerful echoes of John Dunne's disaster dreams – in which Ivon returned but then, his manner secretive and strange, would suddenly be 'lost to me again'.

It was evidently during his grieving period that the idea of making and flying an aeroplane took hold – though his interest had first been aroused years earlier, on a lovely summer's afternoon in Nuneaton. Not far from the vicarage a fete was being held. He and Ivon were sitting at a window, listening to a brass band, when 'Suddenly a great balloon rose from the town, climbing steadily and fast and high into the sky.' The brothers were so captivated they talked about building a flying machine, drawing most of their ideas from Jules Verne's *Clipper of the Clouds*. The Clipper had needed at least a dozen vertical fans to get airborne, so vertical lift became their chosen method of propulsion and, for a time, so absorbed them that they even wrote to various 'makers of electric fans to ask about the thrust and horsepower required'.

But it wasn't until 1907 that Geoffrey, earning forty-five shillings a week as a draughtsman at the Motor Omnibus Construction Company of Walthamstow and driven half mad by tedium, decided that aviation was something to which 'I was prepared to give my life'. At that point he'd never even seen an aeroplane, but appears to have followed a series of thrilling demonstrations given by Wilbur Wright at Le Mans and widely reported in the British press. He called on his grandfather who, ever since he could remember, had been bankrolling the spendthrift de Havillands. Jason Saunders poured them whiskies and talked for a while about a Mr Morris (the future Lord Nuffield) who owned an Oxford motorcycle shop and hoped to run motor buses in place of the horse-drawn trams that had helped Saunders make his fortune. When Geoffrey outlined his plans he said, 'You really think you know enough about it to build a flying machine?'

Geoffrey said, 'Yes, I'm certain I do, and I've an overwhelming desire to fly.'

The old man pondered the end of his cigar. Then, quite unexpectedly, he said, 'I intended leaving you a thousand pounds in my will, but if you prefer to have it now you can.' He added, 'But there will be no more later on.'

It was a huge sum, and his grandson, stunned by his munificence, hurried back to London. There he called at the digs of a friend, 'a tall and good looking Cornishman' named Frank Hearle who was doing 'uncongenial' work at a bus garage in Dalston. He told Hearle he was going to design an aeroplane engine and construct the airframe it would lift. 'Will you come and help me?'

'Hearle gave a look of pleased surprise. "I'd love to," he said.'

Hearle was receiving two pounds ten shillings a week from the bus company yet, instantly, he agreed to one pound ten shillings from Geoffrey's more constrained budget. In 1908 the two moved into a Kensington flat, and persuaded Geoffrey's sister, Ione, to keep house for them. She was an amusing, argumentative, strong-minded Socialist – 'more to the Left than the most rabid Labourite' grumbled Geoffrey from his own position out on the more rarefied extremities of the Tory Right – yet he was close to her, and delighted when she and Hearle fell in love and married. The plane, though, always came first.

At Crux Easton he had befriended the son of the rector from the next village. Gurth Churchill was a partner in an estate agency with its offices in Bedford Court Mansions, Bedford Square. Geoffrey asked if he could rent a room in which to design his engine, and 'generous terms' were agreed. Next he needed premises in which to actually build his aeroplane. He and Hearle found a shed in Bothwell Street, Fulham, which was available for a pound a week. For under £20 they equipped it with 'planes, saws, chisels, files, gluepot, spoke-shave, vice, drill, etc.' The engine, taking shape on a Bedford Court drawing board, would be a 50hp water-cooled 'four cylinder opposed' or 'flat four' with ball bearings for big ends. When it was complete he took the drawings and tracings to Ivon's old firm, the Iris Car Company of Willesden, who agreed to build it for a mere £220. ('I tried not to show pleased surprise in accepting the figure.')

No estate agency has premises at Bedford Court Mansions today. A white-haired porter, out on the steps enjoying the sunshine, looked mystified when I mentioned it. 'News to me. But some of our older residents might remember.'

'They can go back ninety-nine years?'

'Ah. That long ago, eh?' Well, that made it history and, as it happened, he had some interesting historical objects right there: a row of elegant brass-ringed circles set in the pavement. As he explained that they were, in fact, the old coal holes preserved at great expense, I surveyed the place. The Mansions are a fashionable apartment building, five storeys of mellow red brick occupying a city block next to the British Museum in Bloomsbury, its leaseholders a cosmopolitan crowd who, irrespective of colour, race or creed, have about them the smell of money – which, oddly, as they went about their business that day, seemed to be giving off quite definite floral notes: orange blossom, jasmine, lily of the valley (and that was just the men).

Two young, deeply tanned women, both dressed to the nines and sporting fabulous hats, came staggering out the door overcome with laughter. '*Telephone* sex?' whooped one. 'With *Tallulah*? You've got to be kidding!'

'Morning, ladies!' said the porter, saluting smartly. 'Off to the races again, are we?' But they ignored him.

To me he said, 'Look, I've just had a thought. We don't have an estate agent here, but we do have some quantity surveyors. Roughly the same line of business, isn't it? They're at the west end of the Mansions. Number 12. In Adeline Place.'

I thanked him, and headed off, found No. 12, on the ground floor with its own private entrance – and knew, with an odd kind of certainty, that these had been Gurth Churchill's premises.

As I approached it the door opened and a man ran down

the steps. In his early twenties, lanky and blond, face fashionably stubbled, he was hurriedly shoving papers into an expensive pigskin briefcase.

'Excuse me,' I said, 'but is this your office?'

He paused and stared. 'What?'

I explained, in a sentence or two, what had brought me here. He thought I was mad. '*Aeroplanes?*' he said. 'You've lost me. De Havilland make shirts. Bloody good ones, too. I've got some. Moss Bros stock them.'

I also had a few, but somehow doubted that Geoffrey's family was involved. 'OK, but I don't think . . .'

He wanted to be rid of me.

'Now isn't a good time to visit,' he said. 'Everyone's *furiously* busy. You could call up for an appointment but, honestly, there's nothing to see. Now, if you don't mind . . .'

And with that he was off, yelling for a taxi. I made for Tottenham Court Road, the nearest Underground (Geoffrey also used to come here by Tube), and never went back.

I wouldn't have minded having an apartment at the Mansions, though. Bedford Square, for which I have a long-standing affection (my first books were published, by an old family-run firm, at No. 47), lay just across the road, a large, beautiful, tree-shaded space in an unusually interesting area of London. There were, if you could afford the prices, worse places to live.

Walking along Bothwell Street, Fulham, however, I couldn't for the life of me work out where Geoffrey had kept his workshop. It was a small, shabby throughway lined mostly by dowdy town houses, many with multiple entry buzzers signalling conversions into flats, but featuring too a couple of nondescript apartment blocks. A forest of For Sale signs put up by Sebastian Estates ('the Fulham property specialists') indicated that that Bothwell Street's population might be restless and transitory.

It was a fine, breezy Saturday afternoon and, after loitering for ten minutes, I still hadn't seen a single living soul. Then a red Mondeo turned in at the northern end and stopped. The driver went to a house and rang all four doorbells. No one came. Seeing me, he said, 'JB Cars. Are you Mr Pugh?'

He was an elderly Indian, quite stout, with bushy silvery eyebrows and pomaded silvery hair. As he sighed and glanced at his watch I asked if he'd lived in the area long.

'Me?' he said. 'Fifty-two years. Actually, I went to school just half a mile away.'

He was just the person I needed. 'Do you remember, by any chance, ever having seen a shed here? In Bothwell Street?'

He stared. 'A *shed*?'

'Yes.'

'You mean, like a garden shed?'

'No, a shed big enough to build an aeroplane in.'

That made him smile. 'An aeroplane? How could you fit an aeroplane into Bothwell Street?'

'Just a small one,' I said. 'Not a jumbo.'

'Maybe, though, you could fit Bothwell Street into a jumbo.'

I laughed and told him, briefly, about Geoffrey. When I mentioned that the finished machines had been moved to Hampshire, he said, 'Fulham has many rich people, you know. Maybe not in this part, but they are here. And they like it because at the weekend they can easily get to their country houses in places such as Hampshire.'

The Great West Road! And suddenly, as he left to ring more doorbells, I understood why this curious little street had been chosen. It was already on the way to Beacon Hill.

## Chapter Seven

# The Patriarch of Stag Lane

Geoffrey prepares to depart from a field in Aldershot – having first,
presumably, checked it for larks' nests.

Geoffrey had read Maxim's book on aerodynamics, but the machine
taking shape in Bothwell Street came straight from his head. 'The
main and dominating idea was to build something quickly and not
too novel, because I felt the only way to learn was by experience,
whether or not this ended in disaster.' He made few drawings but
worked intuitively, arranging various complementary planes and
surfaces like a sculptor, getting the look, feel and balance right. The

fuselage was a mere girder; everything else, like the fourteen-foot elevators – great supplementary wings rigged over the nose wheel – would provide lift.

During this period he married Louie Thomas, governess to his sisters at Crux Easton – without fuss, it barely gets a mention in his book – and now, daily, she came to the shed 'to sew the wing fabric and make tea'. (There is a picture of her at work, a handsome woman with large eyes, broad cheekbones and a pensive look. She wears long skirts and a broad-brimmed straw hat, and sits at a Singer sewing machine with folds of wing fabric cascading about her feet.) In the evenings they returned to their flat at 32 Barons Court Road, within easy walking distance of Bothwell Street. A quiet residential thoroughfare, it was tucked away behind the new terracotta-coloured Barons Court Underground station, which had opened just four years earlier.

Today, attached to the station, there's a grocery, and a beauty salon offering such treatments as a £50 'Indian Head full body massage', while Barons Court Road, like Bothwell Street, has a preponderance of flats, with a twenty-four-hour locksmith's number posted by each front door. No. 32 is a spacious four-storeyed London town house evidently fallen on hard times. Badly in need of paint, with grimy, unswept front steps and overflowing plastic dustbins, it boasted a blue plaque – 'Sir Geoffrey de Havilland Aircraft Designer Lived Here 1909–1910' – which, on such a place, looked fanciful, almost funny, like a *Big Issue* seller wearing the Order of the Garter.

It was a frantically busy time for Geoffrey, who soon, for extended periods, would be only here at weekends (puzzling the neighbours with his upper-class features and permanently grease-grimed fingernails). At Bothwell Street, finally, they completed their machine, a boxy biplane with a crankshaft that drove, through bevel gears, two propellers turning in opposite directions. He and Hearle dismantled it, placed it in a hired van and took the Great West Road towards the Hampshire Downs and John Brabazon's sheds at

John Moore-Brabazon, holder of the Royal Aero Club's
Flying Licence No.1, prepares to launch the world's first flying pig
from Shellbeach Aerodrome, Leysdown.

Seven Barrows.* Set on Lord Carnarvon's estate at the foot of Beacon
Hill they had become available (at a cost of £150) when Brabazon
moved his machine down to Eastchurch. The downland, he recalled
wistfully, was 'a lovely and romantic place to go flying for the first
time, and brought back many happy memories of . . . natural history
expeditions as a boy. I walked carefully over the ground we should
be using, heard larks singing high in the air, and was able to find two
nests and mark their positions with thin sticks so we could avoid them
when taxiing.'

* Brabazon, rich, handsome and ambitious, loved being talked about – at
Leysdown he made the world's first ascent with a live pig (in a wastepaper
basket strapped to a strut) and was a natural show-off (after the Aero Club
awarded him its first licence his car registration read FLY 1). Appointed

A friend sold him an old Panhard for £45, at the Carnarvon Arms in Whitway, three miles from the sheds, a widowed landlady, 'one of the kindest of women', gave them, for £3 a week, two bedrooms and a sitting room along with hot breakfasts, hot dinners, even packed lunches. The Carnarvon Arms, a spacious nineteenth-century coaching inn, is still trading, though when I mentioned the widow's terms they caused a certain incredulity. The young duty manager did some rapid sums and said, 'A single now costs £69.95 – that includes a full English breakfast, by the way – so two would be £139.90, multiply that by seven . . .'

'By five,' I said. 'They usually went home at weekends.'

'OK. By five comes to £699.50, then add the hot dinners, with wine say forty quid for both, that's another two hundred, also packed lunches, say twenty. You'd get very little change from £1000 a week. And I don't know about the sitting room, if we had such a thing today we'd want at least another £50 a night for it.'

He showed me one of the rooms, stylishly furnished with a comfortable bed, wide-screen television, tea- and coffee-making facilities and Internet access. 'Thanks to the M4,' he said, 'we're only an hour from London. So we get the tourists coming to see Highclere, of course, but there's also a lot of corporate business. Tomorrow we're doing a stand-up lunch for ninety followed by a sit-down dinner for forty-five.'

My own lunch, of liver and bacon, was eaten at a table overlooking the road down which, early each morning, Geoffrey and Frank set off, passing close to Highclere Castle with its curious Victorian Gothic resemblance to the Palace of Westminster – both were

---

Minister of Aircraft Production in Churchill's wartime government, he gave his name to the biggest British aeroplane ever built. The Bristol Brabazon, its wingspan *thirty-five feet longer than a Boeing 747's*, carried eighty pampered passengers (soft beds in spacious cabins, cocktail lounges, a dining saloon, even a cinema) and was powered by eight engines driving counter-rotating propellers. Yet it proved so hopelessly uneconomic only the prototype ever got built. John Brabazon was still riding the Cresta Run at seventy.

designed by Charles Barry – then halting just a mile or two further on at the Seven Barrows field, site of their sheds.

It was a pleasant existence, working in a beautiful corner of England, hearing only wind and birdsong and occasionally, like pastoral grace notes, the distant tinkle of sheep-bells. They saw few visitors, though one afternoon Brabazon appeared with a youngish, affable man who, it turned out, owned Highclere Castle and all the land as far as the eye could see. George Edward Stanhope Molyneux Herbert, fifth Earl of Carnavon and already a distinguished Egypt-ologist (eleven years later he and Howard Carter would electrify the world by revealing the tomb of Tutankhamun), promised to keep the grass mown. In return he asked that his shooting parties be allowed to shelter in the sheds when it rained. He and Brabazon examined their aeroplane with great interest though, says Geoffrey huffily, 'I doubt whether they seriously considered it would ever fly.'

Winter set in. Arriving in their open Panhard they were unable to move their fingers until they had held their hands, for ten minutes, over a glowing brazier. The first de Havilland was coming along nicely. It took several weeks to fit the gleaming Iris engine – torque tubes protruding from either end of the crankshaft – but then, on a crisp December morning in 1909, they rolled out the completed aeroplane, 'started up the four cylinder motor and watched the twin propellers rotating . . . like great paddles, catching the winter sun on their blades.' That first time it worked perfectly but, when he tried again, 'with a sudden screeching sound a bevel gearbox collapsed', then during taxiing trials they had transmission trouble caused by a twisted driving shaft. And though it attained excellent speeds on the ground it stubbornly refused to fly.

Geoffrey had not helped matters by installing curiously compli-cated controls. He steered with his right hand, moved those giant barn-door elevators (which would take him up or down) with the left, and the ailerons (which moved down but would only take him up) with his feet. They required, says aviation historian Harald Penrose, 'considerable thought to operate', and certainly didn't help a rookie who had never been off the ground before. These complex, contra-dictory controls proved to be an important factor when he finally 'got the first de Havilland machine off on its first flight'.

It lasted, of course, only a second or two, crashing in front of Frank, Hereward and his shaken father.

The wreckage was returned to Fulham. 'Money,' he noted, 'was running low,' so speed and simplicity were now essential. He replaced the twin propellers with a single one mounted direct on the engine shaft – thus disposing of a heavy iron flywheel – and simplified the landing gear, working in a calm, unfussy way that caused Frank to tell a friend, 'I am always amazed at the way de Havilland managed . . . He never gave orders, just talked matters and problems over, and you knew what he wanted and did it. From the beginning he always seemed like an elder brother to me.'

The new machine was taken to Seven Barrows where, after three weeks of scaring the rabbits, all seemed ready. One 'beautiful evening in late summer' Geoffrey made several runs into a light breeze then pulled up beside Hearle. 'I'm going to try one more,' he yelled over the racket of the engine. 'Will you lie down as I go past and watch if you can see any daylight between the wheels and the grass?'

Accelerating towards Hearle – who was prone on his stomach – he reported, 'I was going faster than ever before . . . I must have been travelling at 25 or 30 mph when I passed beside him, and at once eased back, coming to a halt.' Hearle rushed towards him, shouting and waving his arms. 'You flew all right!' he gasped. 'You were several inches off the ground for about twenty yards. Well done!' Geoffrey, somehow aware this would be 'the most important and memorable moment of my life', leaped exultantly from the aircraft (jumped for joy). Later he would capture the world altitude record – climbing to 10,500 feet over Salisbury Plain – yet nothing could match the rapture of a maiden flight made so low the buttercups rocked in his slipstream.

Louie was summoned from Kensington, a spell in the country evidently a reward 'for hundreds of hours on her sewing machine'.

Geoffrey warmly commended her for this, then also noted – in a casual aside – that she was heavily pregnant; 'a few weeks later our son, Geoffrey, was born at Crux Easton.' No further mention was made of mother or child until, after a month or so, they showed up at Seven Barrows for the trials. These were going well. He was venturing higher and higher, completing figure-of-eight manoeuvres, making banked turns, gaining confidence, even one day climbing through heavy cloud and having a transcendental moment 'in a blinding world of billowing whiteness that stretched in every direction, magnificent and vast and thrilling.'

Oddly, in his memoir, there is not a single reference to the press. He was, perhaps, too preoccupied to notice them; they, however, certainly noticed him, and on 22 September 1910 the *Newbury Weekly News* cleared its front page of the usual equine and agricultural items for a genuine scoop. 'THE ALL BRITISH AEROPLANE' it trumpeted. 'The de Havilland Aeroplane Flies at Beacon Hill.' Having dutifully recorded the feelings of Lord Carnarvon – who, preparing for a new expedition to Egypt, had been present and 'elated at the success which attended the efforts of the flying men' – its reporter described the biplane being 'taken from its shed in the early evening when the atmosphere was at its calmest', then, moments later, clambering into the air trailing smoke. 'The speed,' wrote the reporter, 'is great, possibly 30 miles an hour.' He added, 'However, Mr. de Havilland has not yet completely mastered the vagaries of the machine on which a false move means a big smash. He descends to the ground and runs, taking jumps of 40 or 50 ft.' Geoffrey, evidently, was practising rudimentary circuits and bumps, but then, 'just as the light was failing, the biplane was stabled.'

It had been created by a man who 'relied solely on his own inventiveness . . . and his own ideas, and not on advice acquired from others.' The key component was the engine, 'absolutely M. de Havilland's own invention, many details of which he has protected.' All he appears to have told the reporter was that 'it weighed 200 lbs, developed 50 h.p. and drove a double-bladed mahogany propeller.'

Yet Geoffrey mentions none of this. Instead he writes of fitting another chair so that Frank could take his first flight ('we went up to a hundred feet') then of persuading Louie to follow suit. And here we get a rare glimpse into the state of his marriage. Publicly Geoffrey would become known as one of the technocratic geniuses whose work helped define the twentieth century yet, privately, he turned out to be a Victorian patriarch with very fixed ideas about a woman's place. Louie agreed to go up 'because she knew *it would please me* and that she felt it *her duty* [my italics] to show her interest in our work.' His work, though, was so dangerous no insurance company would touch him; not only was he flying a rudimentary home-made machine, he was still learning how to do it – and would carry on even if it all ended, as he himself put it, 'in disaster.'

When Louie went up, amazingly, she brought the baby along too; she travelled over from Crux Easton one afternoon, climbed aboard with young Geoffrey in her arms then clung to him like mad as, sitting in their garden chairs (with not a seat belt between them), the de Havillands flew sedately around Beacon Hill. A picture of them airborne would have made the pages of every newspaper in the country, any publicist worth his salt would have walked through fire for it. Yet no picture was taken, not a single report appeared. It was a family matter. Though Geoffrey guessed that, at eight weeks, his son had 'been the youngest person in the world to go up' he seemed curiously unsure of Louie's feelings. 'I *think* she enjoyed it' he wrote (my italics), but Louie's enjoyment was not the point. Nor were any didactic notions of wifely duty. By her extraordinary act she had demonstrated, quite simply, her absolute trust in him.

Today Crux Easton's school has gone, and so has the pub, its licence transferred to the Greenham Common Air Base. (Eccentrically, the Three Legged Cross never had a bar; beer was poured in the still room, the tankards then brought through to customers by Jack Greenaway, the landlord. Later he became a smallholder, supplying Crux Easton with its milk.) And the telephone exchange – with nine subscribers and a ringing tone consisting of single-level blasts, like a car alarm – has gone also (to London's Science Museum). At the top end of the village stands a red K6 phone box – the classic Gilbert Scott design – that once took penny coins but, when I visited, had a busted door and a malodorous interior; another relic of those days, it, plainly, would soon be gone too.

The church, St Michael's, was unexpectedly small and pretty, a Georgian brick structure with an apse containing a large arched window, some nice panelling around the altar and a fine Italian marble font. A plaque commemorated the Revd Charles de Havilland MA, Rector of Crux Easton for twenty-five years, who died in 1920, and his wife, Alice, who died in 1911. On a pew lay the big leather-bound Oxford Bible from which he had read to his congregation and which, according to a copperplate inscription on the flyleaf, had been 'Presented by the Earl of Carnarvon AD 1882.'*

Opposite the church stood a white cottage badly in need of paint. A man, burly and unshaven, sat reading the *Daily Mail* in a small van parked outside. When I asked if the cottage had, by any chance, been the old rectory he pointed next door to a large property set behind big gates. 'That was the rectory. Long time ago, of course, it's a private residence now, Crux Easton House. Been added to a bit.' *A bit?* I'd never laid eyes on a rectory like this. With its lofty ridge-tiled roof and pedimented entrance facade it looked more like a minor ducal palace (and laid claim, incidentally, to an odd fragment of history: Sir Oswald Mosley, leader of the British Fascists, had been incarcerated here for eighteen months after leaving prison. He was a hoodlum, but an upper-class hoodlum – married first to a Curzon, then a Mitford – so, though the nation happened

---

* A recent entry in the visitors' book was written by a woman whose great-great-great-grandfather, baptized here in 1788, had taken part in the Machine Breaker Riots of 1831 and been deported to Australia. He liked it there so much, however, that, pardoned after seven years, he returned to Crux Easton, collected his family and took them back to Toowoomba, Queensland – where his descendants lived still.

to be at war with his Nazi chums, he was left free to wander, entirely unsupervised, within a three-mile radius of the village). Now, when I went to try the gate, the man said, 'Nobody's in.'

He'd certainly heard of Geoffrey, but spoke of him rather dismissively, alleging that he'd robbed Crux of its able-bodied men. 'They all went off to jobs in his factory. Every last flaming one of them worked for Geoffrey de Havilland.'

The way out of this quiet little village led past the old wind engine. Built in 1894 by a man named John Titt, it had recently been restored yet, on this hot, still morning, its forty-eight spinning canvas sails barely moved. Following the route taken by Louie and her baby in their pony and trap, I set off down a lane wide enough for only one vehicle and, routinely, was forced back to passing places by pushy, baleful-looking country drivers. In a grassy roadside clearing I came upon two gypsy caravans, and a stout Romany woman washing clothes in a plastic bucket. The trees – beeches, limes and oaks growing with almost tropical abandon – often met overhead to form a tunnel, the filtered sunlight then filling it with a soft, virescent luminousness.

I passed Woodcott and Dunley, admired their picture-postcard cottages and gardens, joined the busy A34, moments later came to a lay-by and a notice saying, 'Geoffrey de Havilland First Flight Memorial ← 80 metres.' The arrow pointed along a dried mud track which went winding up an easy wooded gradient: that had to be Beacon Hill. Cross-country power lines crackled noisily overhead as I arrived at a block of granite set on a concrete square and fenced off like a small grave. 'Geoffrey de Havilland,' I read, 'assisted by Frank Hearle, carried out his first flight in his home made aeroplane here at Seven Barrows on 10 September 1910.'

But where had the sheds been? Two of the barrows, or prehistoric burial mounds, loomed some distance away in a meadow of tussocky grass and, looking for shed remnants, I walked around them. Then I noted that the power lines passed through an electricity substation surrounded by tangled scrub and built higher up, thus avoiding problems caused by heavy rain on the valley bottom. The sheds would have had ready-made, packed-earth floors and, though I never found a shred of evidence, I'm convinced they stood where the substation stands now.

On one side of the rising track was a wooded gradient, its steepness increasing incrementally with altitude, on the other a gently sloping barley field. Quite high up, hunting for wild strawberries, I heard larks – perhaps descendants of the ones whose nests he'd so scrupulously marked. As I tried to spot them, a woman came marching towards me. She was in her mid-forties, stocky and dark-haired, wearing a linen hat and walking boots, and carrying a rucksack. 'Hullo there,' she said. 'Would you happen to know where this track goes?'

'Down to the de Havilland Memorial.'

'Oh, good,' she said. 'That means I'm not lost. I'm doing bits of the Wayfarer's Walk, yesterday I was on Watership Down, not far from here. I think I saw Richard Adams's house. Didn't see any flipping rabbits, though, not a single one.'

'He probably made it up,' I said.

She smiled. 'Oh, surely not.' A crested bird flew by. 'Lapwing. There are millions around. Have you noticed? Some interesting little plants, too.' She took a water bottle from her rucksack then, sipping as she went, wandered along the trackside, finding and identifying wild thyme, clustered bellflowers and pink rock roses. 'And there are tiny bees called *Osmia bicolor*. They've got red tails and eat birds-foot trefoil.' She pointed to some low-growing yellow flowers. 'That stuff. And live in abandoned snail shells. They're quite rare – but they like chalk downland and are known to flourish on Beacon Hill.' No miniature bees, however, happened to be in our vicinity and, after we'd chatted for a moment – she was a London GP taking a few days off – she headed one way and I the other.

A beech forest now occupied the steep side of the hill, and I guessed that Geoffrey had made those take-off runs down its next quadrant, the long, gradual slope on which the barley grew. I

imagined that tiny biplane getting airborne over the quiet country
lane that was to become the A34, one of Britain's worst racetrack
roads. Then, stampeding the sheep, he might perhaps have made a
circuit of Beacon Hill, or passed low over the spot on its summit
where, in a private family plot, his friend Lord Carnarvon would one
day be buried. (The fifth Earl was the first victim of Tutankhamun's
famous curse, his body – embalmed by Egyptians using techniques
not dissimilar to those employed three and a half thousand years
earlier on Tutankhamun himself – shipped home in 1923.) It was
very quiet. I lay on a mossy bank under a chestnut tree and, half
hypnotized by the play of light on leaves, dozed off. When I awoke
the first thing I saw was a fox, sitting just yards away, observing me
quietly.

*Flight* magazine, in the course of a piece congratulating Geoffrey
on his progress, noted admiringly that he had not been held back
by 'the restrictive nature of the trial ground'; Seven Barrows, then,
had been pitted with rabbit warrens. Busy refining the machine
(contriving an enclosed cockpit against the English weather) he ran
short of money and dropped a note to Lord Carnarvon, inviting
him to put up £200 of risk capital; but Carnarvon, away in the
Valley of the Kings, never replied. It was a worrying time. 'Frank
and I had no regular jobs to turn to, and my grandfather's thousand pounds was just about exhausted.' Then, at the 1910 Olympia
Show, he bumped into Fred Green, an old engineering friend. 'Still
on the buses?' enquired Fred. Geoffrey told him he'd built his own
aeroplane, but was now unemployed; Fred said he worked at the
Farnborough Factory and might be able to help. At the moment they
were interested only in balloons and airships. 'But I happen to know
the Superintendent is keen to have an aeroplane too. Why not offer
him yours, and ask for a job at the same time?'

The head of the Army Balloon Factory, Mervyn O'Gorman,
lived at Embankment Gardens, Chelsea. A tall, bald, amiable Irish-
man sporting a gold-rimmed monocle and amber cigarette-holder,

he welcomed Geoffrey when he arrived clutching his drawings. Geoffrey said, 'I'm afraid I don't know much about aeroplanes.'

'I shouldn't worry about that,' replied O'Gorman. 'Very few people know anything really reliable about them. It's still mostly a matter of trial and error. At Farnborough we're very keen to start on aeroplanes, but the snag is that the War Office people haven't got much faith in them.' (This astonishing exchange took place as, in Germany, the first combat aircraft were being built, the first pilots trained; and, while Berlin authorized £1,500,000 for aeronautical research, Whitehall wouldn't part with a penny over £20,000.) He added that, if the price was right, he would try and persuade them to buy Geoffrey's machine. So what did he want for it? Geoffrey had no idea. How could you place a value on something like that? The first figure that came into his head was £400, and, months later, the War Office agreed. At the same time they authorized O'Gorman to offer Geoffrey a job designing new machines, and even to create a post for Frank – the only mechanic in the land who understood aeroplanes. Geoffrey, jubilant, offered to repay Jason Saunders at least part of his £1000. But the old man reminded him it had been a gift, adding that anyone who could, on £500 a year, create a brand-new engine, build two aeroplanes, employ an accomplice (who married his sister), get wed himself, raise a son and feed five mouths deserved every penny he could get. Saunders wanted none of it back.

Geoffrey and Louie rented a small house near the Balloon Factory. O'Gorman, on a whim, had also made him chief test pilot, and she was delighted by the financial security they now enjoyed. He, however, found his cavernous new workplace riven by ideological squabbles. Though aeroplanes had been grudgingly accepted, the Factory still clung to the traditions of the lighter-than-air brigade (aviation's self-appointed lords of creation) and he was astounded by the hostility he encountered; when, for example, he and Frank were unloading their machine, 'the balloon men . . . watched in dour silence and made no move to assist us.' No wonder. They surmised, quite correctly, that these oily-fingered, goggle-eyed, fixed-wing intruders would – very soon – make them all redundant.

When the balloonists were, finally, eased out, Geoffrey found himself surrounded by clever, like-minded people with whom he formed enduring friendships. Those he and Louie invited home for dinner included Sam Cody, the brilliantly innovative J. W. Dunne, Alliott Verdon Roe, an aeronautical genius who, like Geoffrey, went on to build a great company, Harold Bolas, a specialist in the mathematics of structural stressing ('All problems,' he liked to say, 'are soluble in beer'), and the scholarly, likeable Major Frederick Sykes, Commandant of the Royal Flying Corps, with whom, over Salisbury Plain, he achieved his 10,500 foot altitude record. (Having almost frozen to death – they'd been two miles high – and getting hopelessly lost on the way down, they blundered around Hampshire following the smoke from country trains. One, surely, would take them to a familiar station, and it became a kind of game; eventually, after pursuing various branch lines past a series of baffling pastoral junctions and dead-end rural halts – indeed, having zoomed over half the county – they finally made it home.) Sykes* became Governor of Bombay and, in memory of that day, his widow asked Geoffrey to strew his ashes over Salisbury Plain.

Some of the early Farnborough-built de Havillands contained design flaws – one awful consequence being the death of Edward Busk, a 'round-faced and smiling' scientist, who had been at King's College, Cambridge, with the poet Rupert Brooke and, whenever he came to the house, sang out-of-tune songs to their new baby, Peter. One day Busk's engine inexplicably caught fire and Geoffrey, flying close behind, watched him spin into the ground 'like a flaming torch'. (For many years the scene kept recurring in dreams.) Geoffrey

---

* The journalist C. G. Grey claimed that Sykes, when young, often went flitting around Asia on mysterious official errands. Once he walked three and a half thousand miles from Beijing to Simla, the Himalayan retreat where India's viceroy, and his spies, spent their summers.

himself crashed just once, breaking a jaw and losing several teeth – recovered by a kindly mechanic who brought them to his hospital bed in an envelope. And, while he worked furiously to make his own machines safe, he was known also as someone prepared to tackle problems others found intractable. When, for example, the Duke of Westminster presented the Factory with the fragmented remains of a biplane that had evidently collided with a cliff, Geoffrey turned them into a long-range military machine so quiet the press, fascinated, came to regard it as a kind of early stealth bomber.

He became – it is believed – the first Englishman to loop the loop. (Safety harnesses had not yet been thought of and allegedly, when an unnamed Royal Flying Corps pilot tried it with a companion in the rear seat, he glanced back to find the seat vacant.)

In 1911, as the last of the lighter-than-air men departed, O'Gorman gave his vast shed a more appropriate name; the Balloon Factory became the Royal Aircraft Factory. And he gave Geoffrey a new title, Inspector of Aeroplanes; as teams of designers and engineers converged on Farnborough, he test-flew every new machine produced there. Like many civilian pilots he had joined the RFC but, due to his fractured jaw, was judged fit only for home duties. So, when hostilities broke out, Lieutenant de Havilland was sent north, given an antique 40 mph Bleriot – without guns, bombs or wireless – and told to look for U-boats between Aberdeen and the Firth of Forth. He existed in a state of almost catatonic boredom until the War Office wondered why someone so crucial to their effort was spending his days crawling up and down a stretch of empty Scottish coast. They made him a captain and brought him home.

He'd already left Farnborough. In 1914, wearied by the bureaucratic drudgery his Inspectorate required, he joined the Airco factory at Hendon. There he received a royalty payment for every plane built to a de Havilland design and, as the war progressed, grew rich. Though he claimed to find the subject of money 'irksome' he soon had enough to buy a big house in Edgware where, as the war ended, Louie gave birth to their third son, John. Then, to everyone's astonishment, Geoffrey suffered a nervous breakdown. How could this be? What had plunged the famously serene Geoffrey de Havilland into a state of 'almost continuous' despair? Well, there may have been a hereditary element: his mother, after all, had endured her own breakdown in Nuneaton. And the risky nature of his working day was certainly a factor; also there were the dreams: Ivon's death, Busk perishing in a fireball. He may, too, have had more universal worries. Aeroplanes were transforming the nature of global politics; when a Shorts S27 biplane, back in 1912, was launched from a battleship (HMS *Africa*, off the Isle of Sheppey), the Admiralty began talking about carriers, and how even the most distant places would fall within their operational ambit. Gone for ever was the secure, predictable, God-fearing, lamp-lit, horse-and-buggy world into which he had been born. His 'nerve specialist' prescribed several months' rest in a nursing home.

Afterwards he bought a camera and made a film about the swallowtail butterfly which, bought by a movie company, was shown in cinemas throughout the land.

He built a cottage at Crux Easton and, at weekends, flew the family down in a Puss Moth, landing (avoiding the larks) in a field behind the church – which, when I saw it, was a tangle of summer wildflowers and long bleached grass. Occasionally he took the villagers for joyrides: 'cloud-hopping', they called it.

In 1920 he quit Airco to found the de Havilland Aircraft Company, buying an aerodrome at Stag Lane, near Edgware, from two ex-pilots who were using its hangars for a chocolate-making venture. (It failed.) Here, having appointed Frank Hearle his works manager, he produced the Gypsy Moth, which, redesigned as a trainer for the RAF in 1931, became the Tiger Moth. Both, today, are regarded as masterpieces, the latter having become one of the most famous aeroplanes of all time.

At Stag Lane he started a flying school which, thanks to the patronage of the Duchess of Bedford, became popular with the aristocracy. (Though, supposedly, she died in a violent storm, the Duke

Geoffrey and Louie, probably at Stag Lane, date unknown.

always believed that, depressed by the onset of old age, she had crashed her Moth deliberately.)

When a quietly spoken Oxford engineering graduate named Norway applied for a job at Stag Lane, Geoffrey, impressed by his knowledge of aerodynamics, hired him on the spot. Some years later Nevil Shute Norway left the aviation industry to became a writer. *A Town Like Alice* may have been his best-known novel, but *No Highway* was the one that interested Geoffrey. It was about a brand-new, high-speed passenger plane brought down by metal fatigue – and foretold precisely what would happen to the de Havilland Comet. In January 1954, after two years' faultless service, one crashed into the sea off Elba then, just three months later, another went down off Naples. A form of vibration known as flutter had caused the structural failure in Shute's fictional machine – just as it would in Geoffrey's real ones. Looking back on his ex-employee's extraordinary prescience he wryly recalled that, 'Norway had always been a keen student of "flutter".'*

Then – as now – Stag Lane could be reached by catching the Northern Line to Burnt Oak, just south of Edgware. There I found a busy town whose citizenry was mostly black and Asian – reminiscent, say, of Suva, Fiji, or Port of Spain, Trinidad. There were, of course, white faces too, yet I sensed that, some day, the whiteness would be bred out of them, that this corner of Britain would be wholly colonized by people – or descendants of people – from places that had themselves once been colonies ('Chickens come home to roost,' as a young neurosurgeon, London-born of Sri Lankan parents, murmured to me). Markets selling comestibles never seen in Sainsbury's traded beside Sly Afro-Caribbean Foods, Kashmir Halal Meat and Kabul Gate Seafood, while the likes of Hassan Mens Outfitter, Ebony Hair & Beauty and Shu Bizz hummed with activity.

On the western side of Burnt Oak Broadway I asked an Indian postman for directions and found myself following a sign reading 'Gurdwara Brent Sikh Centre' which, in fact, sent me straight down Stag Lane. Lined with small, well-kept houses, it led to a junction where, suddenly, the street signs bore names I knew, the first two commemorating a couple, Jim Mollison and Amy Johnson, whose

* By testing a Comet fuselage to destruction scientists at Farnborough's Royal Aircraft Establishment (the old Balloon Factory) discovered that metal fatigue lay behind the disasters. Design changes were made, and the Comet re-established so emphatically that, almost *sixty years later*, the Nimrod spy-in-the-sky military version still remains in service with the RAF. (Yet the delay cost Geoffrey untold millions in sales. The Comet, being first, had served to whet people's appetite for jet-propelled air travel. It was Boeing, seizing the moment and cornering the market with their 707, who satisfied it.)

tempestuous relationship and remarkable exploits had, long ago, fascinated the British public. On the corner of Mollison Way, which sloped off to the right, was a small four-storeyed residential complex set in a shadowy garden: Amy Johnson Court.

Amy, a slim, pretty secretary – she worked for a Hull solicitor – had been one of Geoffrey's outstanding flying-school graduates, and in 1930 flew solo to Australia in her bottle-green Gypsy Moth.* No woman had ever done that before and, when she arrived back at Croydon, the police were barely able to contain a two-hundred-thousand-strong throng. Next evening a welcome-home dinner was held at the Savoy. Among the celebrities tucking into 'Œufspochés Port Darwin' were Evelyn Waugh, J. B. Priestley, Noël Coward, Ivor Novello, Cecil Beaton, Alfred Hitchcock and, of course, her mentor Geoffrey de Havilland. Later she married Mollison, a handsome, charismatic Scot who, having drifted through his early years – beachcombing in Tahiti, a life-guard on Bondi – became an immensely glamorous pioneer aviator. He too had been a close friend of Geoffrey's.

Another was Sir Alan Cobham, author of the bestselling *Twenty Thousand Miles in a Flying Boat* and a regular visitor to Stag Lane. Cobham, who had flown a Gypsy Moth to Zurich and back (thus becoming the first pilot to cover a thousand miles in a single day), was chronically hyperactive. When, once, he dropped in for lunch, Geoffrey 'watched in fascination as, talking constantly, he put whole potatoes and chunks of meat and bread in his mouth . . . He was still swallowing, and still talking, as he raced for the aerodrome, leapt into his aircraft and tore off.'

I couldn't find Cobham Close. The map placed it on the far side of Mollison Way, but a new development still under construction had, for the moment, blocked access. De Havilland Road was there, however and, at 224–234, a modest brick apartment house with

---

* Named 'Jason', displayed in London's Science Museum, it retains its original green paint and charming registration, G-AAAH.

bicycles in the lobby and, over the front door, a silvery cut-out aero-plane, I spotted a slate plaque saying 'Site of Stag Lane Aerodrome founded in 1916 by the L and P Aviation Co' then, from 1920, 'Home to de Havilland Aircraft & Engine Companies'.

A stocky, middle-aged man approached, hobbling along on sticks. I explained my interest, and asked if he happened to know where the airstrip had begun and ended. He said, 'Uh-oh, I think you better talk to my brother,' then took a mobile phone from his pocket, spoke a few words, and handed it to me. 'Go ahead.' The person on the other end had a rumbling bass voice, and seemed utterly unsurprised by this sudden encounter with a perfect stranger. 'I believe the planes started their take-off run roughly where you and Isaac are standing now, then got airborne round Tennyson and Milton Avenues.' He chuckled. 'Got dead pilots one end, dead poets the other.'

He chatted on, said it had been the unveiling of the plaque in 2001 that aroused his interest in Stag Lane's history. Did I know the first jump with a modern parachute – the kind where a small primary chute pulls the big one from the bag – had been made here in 1925? And that the chief instructor at Geoffrey's flying school had been a Captain Valentine Baker who, later, teamed up with a James Martin to invent the Martin Baker ejector seat? 'They saved the lives of countless aircrew. Thousands of 'em.'

It was all news to me but, feeling awkward about taking up this man's time, and using his brother's phone, I thanked him then ended the call. Isaac, in no hurry, told me his brother worked in a Golders Green convenience store, while he himself was a PE teacher. 'But not at the moment, I've broken a toe, hurts like hell.' He pointed at my notebook. Was I an aviation writer? I said my field was travel. Travel, eh? So had I been to Ghana? No? What a shame! It was where his family came from. The only work his grandfather, an educated man, a *headmaster*, had been able to find in the UK was grave-digging, yet he never regretted the move. 'You know what he used to say?'

I shook my head.

'The streets of London are paved not with gold – but *possibility*.'

When Geoffrey arrived in 1920 just a single house overlooked the aerodrome; ten years later hundreds of bungalows lapped at its perimeter. He and his sons found new premises by flying along country roads and seeing where they went; one, eventually, led to Hatfield in Hertfordshire, which had all the space in the world. To ensure it stayed that way he bought the land adjoining and stocked it with pigs and pedigree cattle; the fact that his farm made money added to the pleasure it gave him. (On weekends he invited friends over to shoot the hares, rooks, partridges, pheasants, even golden plovers that shared his grassy domain.)

Here he created airliners like the Dragon – actually commissioned by the Iraqi Air Force, but also used by Qantas and Imperial Airways – and, in 1934, the Rapide, a pretty but remarkably rugged biplane which saw service all over the world. (As a child I flew in one operated by Fiji Airways; it had a leaking roof, and passengers were provided with umbrellas in case of encounters with rain-bearing cloud.) In 1934 he also produced the elegant, high-speed Comet, designed specifically for the England to Australia race and not to be confused with the jet airliner, unveiled in 1949.

The de Havilland Mosquito was an all-wood, high-speed bomber first flown – by Geoffrey Jnr. – in November 1940. Entering service with the RAF it became one of the fastest (20 mph quicker than a Spitfire), most versatile aircraft of the war. Almost eight thousand were built, providing work for seventy-five thousand people, including carpenters and cabinetmakers. In 1944 his son John, piloting a Mosquito, was involved in a mid-air collision from which there were no survivors. Two years later Geoffrey Jnr. (who made his maiden flight in Louie's arms around Beacon Hill) was at the controls of an experimental jet, the DH 108, which, probably due to excessive stress loads as it approached supersonic speed, broke up over Egypt Bay in the Thames Estuary. (The DH 108's

De Havilland's sons, from left, Peter, John and Geoffrey Jnr.
(who test-flew the first Mosquito, later died at the controls of
an experimental jet). John also died in a flying accident.

sister ship then became the first British aircraft to break the sound
barrier.) Lady Louie died not long afterwards. Cancer was given as
the cause, but everyone knew that what really killed her was the loss
of her boys. Peter, the surviving son, sensibly gave up flying and went
into sales.

In 1946 the famously versatile Vampire was unveiled. The first
jet fighter to fly trans-Atlantic, it was built under licence around the
world (I saw my first as a schoolboy in Australia, forty years later
witnessed Vampires dogfighting over the Alps in the colours of the
Swiss Air Force).

In 1960 the firm was taken over by Hawker Siddeley Aviation.
The airfield from which the Vampire, Mosquito and Comet airliner
first flew, and the industrial complex where seven thousand full-time

employees once worked, now form – following Geoffrey's express wish – the de Havilland Campus at the University of Hertfordshire.

He remarried and spent a good deal of time with Joan in Africa, filming and photographing wildlife. In 1944 he was knighted, in 1962 awarded the Order of Merit. He died in a Watford hospital in 1965, remembered by Harald Penrose as 'an aloof, but unpretentious, loveable personality . . . who rose to every occasion, despite deep sorrows.'

Following a request made in his will, the ashes of Captain Sir Geoffrey de Havilland OM were flown to the Hampshire Downs in a pretty little de Havilland Dove. As it circled, precisely, over Beacon Hill, they were eased out by respectful men and swirled buoyantly off in the slipstream.

## Chapter Eight

# The Airborne Cowboy

Samuel Franklin Cody

Probably the best place to start the story of Sam Cody's extraordinary life is the point at which it was formally ended: that famous country funeral.

His early years in Britain had been controversial. While some saw him as an engaging young fellow with imagination and a distinctly charismatic personality, to others he was just a swaggering, self-promoting, mendacious American loudmouth. Then, as the years passed, all that changed. He charmed people in such a way that the King himself not only sent the pipes and muffled drums of the Black Watch to accompany his coffin, but ordered that it be borne on a gun carriage and draped in the Union flag. Ministers of the Crown, peers of the realm, generals, admirals, MPs, Whitehall mandarins, Fleet Street editors, stars of the London stage, airmen from across Europe, his family and hordes of his friends formed a procession over a mile long. The general public came too (arriving in special trains, overwhelming the little garrison town of Aldershot); a hundred thousand lined the coffin's route from Cody's home, fifty thousand more waited at the military cemetery. Afterwards, beneath a sweet-smelling cordillera of wreaths, the talk was all about the Yankee civilian they'd just buried among the British soldiers.

There is a photograph of him taken a couple of years earlier, Cody the dandy in a well-cut suit, high starched collar and elegant silk tie. He's looking a million dollars, all ready, it seems, for lunch at the Ritz – except that he's also wearing an utterly incongruous Buck Rogers-style space helmet. Since it's visorless we can see the fussily trimmed goatee beard, and the way his exquisite moustache, waxed daily and twirled constantly, projects at least three inches from either side of a smooth, broad, well-fed face (making him look rather feline, like a long-whiskered cat). But then one notes the eyes, which are remarkable: hooded, slightly narrowed, pale green or blue (it's a black-and-white shot) gazing intently past the camera. He may have been vain, but he was also exceptionally handsome.

What people always remarked on, though, was his physical strength. The journalist C. G. Grey watched him 'tying knots in 8-gauge high-tensile bracing wire with his bare fingers', while Geoffrey de Havilland was struck too by the immense power in those hands. There was an assumption, after his death, that he must have been a giant yet, in manner and appearance, was said to have resembled a 'large-sized Peter Pan'. Grey also recalled various eccentricities. He liked giving playful names to things – a magneto was a 'magnuisance', the mysterious power available at the press of a switch 'electrickery' – and, at the Military Trials, used his lasso to rope hapless passers-by, bringing them down like steers. It was intended to be a joke, but few found it funny. And he shared with his friend Wilbur Wright an odd characteristic: neither man could whistle.

For the details of Cody's early life we were, until fairly recently, dependent on his biographer, G. A. Broomfield, a railway booking clerk who had known Cody and admired him unconditionally. 'No one,' he sighs (almost girlishly), 'was ever as completely happy as I was when cycling home with him, or riding behind his trap or motor-car.' Published in 1953, forty years after Cody's death, *Pioneer of the Air* offered the reader excitement, adventure, glamour, celebrities, and *not a whiff of impropriety*: it was the book his family wanted.

They certainly didn't want the world knowing that many claims regarding his early years were false, or that later he had behaved in a way that was scandalously unprincipled.

Broomfield, naturally, offers no hint of this. His story starts during the 1870s when a Chinese cook – whose identity remains a mystery – fetched up on the family ranch in Birdville, Texas, and initiated young Sam Cody into the art of kite-flying. Unaware he was planting a seed that, years later in faraway England, would bloom with spectacular results, he taught him not just basic techniques, but also told him kite stories, for example how, in 206 BC, a Korean general, Han Sing, dangled a lantern from a kite that soared so high his enemies thought he had created a new star and, terrified, ran from the battlefield. The cook, to entertain Sam, may also have constructed musical kites with perforated reeds that wailed in the wind, or tightly stretched strings that twanged like Asian spike fiddles. But, above all, he taught him how to set a kite in the sky and make it stay there.

That is not true.

Broomfield tells us that Samuel Franklin Cody was born, of Dutch and Northern Irish stock, the youngest of three sons, in Birdville, Tarrant County, on 6 March 1861. On the Cody property they grew grapes for raisins.

That isn't true either.

Birdville, at the time, was a bustling riverside community with four general stores, two churches, two schools – one for blacks, one for whites – a ferry boat landing, and a post office run by the first postmistress in America: fifteen-year-old Alice Barkley. It had such a future, indeed, that when an election was held to choose the county seat in 1856, Birdville lost to Fort Worth by only seven votes. But then the river dried up, in 1906 the post office closed, people drifted away, today it's just another Texan ghost town.

That is true.

Broomfield says that Cody, severely injured as a teenager when his home was attacked by Marawnee Indians, crawled for eight miles through woods to Fort Worth. (His family escaped in another direction and, for several years, each thought the other was dead.)

That is not true.

He found work with buffalo hunters who supplied meat to the gangs laying Texas's railways, and discovered he had a gift for horsemanship. He also proved to be such a crack shot people began comparing him with his namesake, William Frederick Cody, alias Buffalo Bill. But then, tiring of the slaughter – 'Cody shoots 'em,' said one contracted butcher, 'faster 'n we can skin 'em' – he tamed horses for the US Army, in 1881 led a drive of 3,275 cattle from Wheeler County, Texas, to Custer County, Montana, a distance of thirteen hundred miles; despite marauding Indians, rattlesnakes, wolves and grizzlies, he lost only seventy-four head. A year spent prospecting in the Klondike yielded little and, in 1888, faced by falling beef prices and a declining cattle industry, he brought his formidable riding and shooting skills to the Wild West shows that had become all the rage in the east.

Much of that is true.

Broomfield now asserts he enjoyed such success as a performer he was able to give up the stage and begin dealing in horseflesh. Also that, on a business trip to England, he met John Blackburne Davis, a fashionable figure who supplied bloodstock to Edward VII and, as a sideline, imported the big black Belgian horses that British undertakers hitched to their hearses. Davis always drove a four-in-hand to the Epsom Derby, and rode in Rotten Row. So did his beautiful daughter, Lela. Cody, out riding himself one day, glimpsed her, fell hopelessly in love and, after a brief, turbulent courtship, proposed marriage. She accepted at once.

But Cody was married already.

Mrs Jean Roberts, of Farnborough, Hampshire, has spent many years investigating the Cody legend, and her version of events goes like this: he did indeed perform in Wild West shows and, after touring through five states with Adam D. Forepaugh, a well-known impresario, won a part in *Deadwood Dick, or the Sunbeam of the Sierras*, a new melodrama produced by Annie 'Get Your Gun' Oakley. Miss Maud Lee, a young actress from Pennsylvania, happened to be a fellow cast-member. He adored pretty women and, having got the consent of Maud's father (possibly by pretending he was related to Buffalo Bill), he wed her and brought her to England, where Wild West shows were also playing to full houses.

*Wild West Burlesque* at London's Olympia featured Sam shooting at glass balls strung about Maud's person – but with the pistol aimed backwards, between his legs. (Maud was content to let Cody do the fancy stuff since once, back in the States, she'd practised her marksmanship in public and a member of the audience sustained severe gunshot injuries.) The show was closed by Buffalo Bill's lawyers, who claimed the term 'Wild West' belonged exclusively to him. So the couple went freelance, cheekily billing themselves as 'Captain Cody and Miss Cody, Buffalo Bill's son and daughter'. When the lawyers, unamused, came calling again, they retreated abjectly.

He did indeed, around this time, meet Lela Blackburne Davis – but she, according to Jean Roberts, was also married to someone else.

None of the above is mentioned by Broomfield. Maud has been excised from the story as effortlessly as Lela is brought into it. All at once she is sharing 'his joys and sorrows', the two of them 'a magnificent example of a true love match.' And how did Broomfield see her? 'A magnificent horsewoman with nerves of steel, Mrs Cody was, like her husband, a splendid shot with both gun and pistol . . .' Also, he says, 'she bore him three sons, Leon and Vivian, born in America, and Frank, born in Switzerland.'

Much of that, according to Jean Roberts, is just plain fiction.

When she and her husband, John, moved into 'Pinehurst' – a roomy late-Victorian red-brick villa on Mytchett Road, Farnborough – the estate agent said it had once been the home of Samuel Cody, who flew the first dirigible in Britain. He pointed to a big bay window. 'Apparently his children hung out there and waved to him as it went by.'

Mrs Roberts, a busy woman with six children of her own, had never heard of Samuel Cody and gave the matter little thought. But then one of her sons, who worked at the Royal Aircraft Establishment, reported that Cody had, in fact, attained a status there that was almost iconic; people still talked about him, and, at home over dinner, he would tell her what they'd said. Gradually she grew interested. (This man had, after all, once lived under *her* roof.) Yet the deeper she delved into his story, the stranger it grew. There were weird inconsistencies. Whole stretches of his early life, she came to realize, had been invented – name, date and place of birth, adolescent years, parents' identity – and others, such as his marriage, simply suppressed. She wrote scores of letters, made hundreds of phone calls, began creating a genuine curriculum vitae for Cody in the years before he became a household name and his every move was recorded by the press.

These days she is a world authority (some would say *the* world authority) on Sam Cody. But, when I called to make an appointment, she said, 'Oh, I'm just a housewife, come any time you want, I'm always here.'

At Farnborough station I gave a cabbie her home address, adding, 'It used to be Sam Cody's place.' He said, 'I know it,' and, at the corner of Lysons Avenue and Frimley Road, halted by a derelict-looking property with boarded-up windows, an overgrown garden and a sad air of abandonment; nobody, plainly, had lived in it for ages. But when I pointed this out he said, 'See that blue plaque? Does that say Samuel Cody lived here, or what?' Undeniably it did, but I made him drive on until, arriving at the Roberts residence, we spotted a second plaque. He muttered exasperatedly, and I felt his peevishness had a point. Where else in the kingdom would one find two blue plaques commemorating the same man in the selfsame stretch of road?

When I knocked on Cody's front door it was opened by a handsome amused-looking silver-haired woman, who shrugged when told of the confusion; it happened. He'd lived here, Jean Roberts said, for five years before moving there, to 'Vale Croft'. And two years later he was dead.

Her husband, John, emerged from his study. Friendly and quietly spoken, he was an acknowledged expert on battleships, and wrote highly regarded books about them. Now, though, having made me cheerily welcome, he went off to brew me some coffee. At the front of the house, overlooking Mytchett Road, a smallish, comfortable lounge contained a mass of Cody material: books, papers, a vast collection of pictures all neatly filed and classified and, most extraordinary of all, an intricately detailed account of his life from birth to death, almost a hundred closely typed pages, each page sheathed in transparent plastic – a labour of scholarship that took seven years.

It was she who discovered that Cody's real name was Cowdery, and that he had been born in the drearily unglamorous city of Davenport, Iowa, six years later than the date he always gave. (It's a mystery why Cody, with the whole of the US to choose from, set his early history in Birdville. Perhaps the name appealed to him, or maybe he'd visited during his saddle-tramp days, and liked it because authentic American heroes were supposed to have had their cradles rocked in such dusty, out-of-the-way places.) She established that he had never divorced Maud and never married Lela; Lela, indeed, was already married to a man named Edward King. A licensed victualler from Chelsea, he had fathered four children – Edward, Leon and Vivian, and a girl, Liese. Lela, who became Cody's lover, or common-law wife, was fifteen years older than him, while Liese, who became, in effect, his adopted daughter, was just six years younger.

I imagined him taking on four kids and a woman who could, almost, have belonged to his mother's generation; he had plainly adored her. Maud, though, had been summarily packed off back to the States where, according to one account, she fell out of a balloon and hurt her head. (Jean Roberts doubts the veracity of that.) She did, however, become addicted to cocaine, was declared mad and locked up in the Norristown Hospital for the Insane, where she spent the rest of her life – though judged well enough to do waitressing work in the asylum dining room. At some point, to feed her drug habit, she stole a pair of trousers. When a journalist claimed the thief had been the legendary Annie Oakley the story rampaged across America, and an outraged Annie sued about fifty newspapers – their files became, for Jean Roberts, valuable sources. Then, after Cody died without having made a will, a lawyer acting for Maud filed a claim on his estate. That also earned press coverage she found helpful.

'Did Lela and Cody have any children of their own?'

'Just one. A son, Frank, who died while serving as a Royal Flying Corps pilot during the First World War, and is buried in Belgium. *His* son, Leslie, until he passed on, lived in Staines.'

I knew there were others named Cody who'd learned, out of the blue, that they hadn't a drop of Cody blood in their veins. It was she, realizing they were, in fact, descended from a Chelsea publican, who had to break the news. It gave her no pleasure. 'I simply handed them copies of my research and pointed out that, at the end of the day, it made absolutely no difference to their Cody connection.'

Cody, from the start, got along exceptionally well with the boys. He now regarded himself as a full-time entertainer, and they became part of a new act – 'S. F. Cody and Family, Champion Shooters of

America' – in which Lela was hung with glass balls; unlike Maud, however, she wore crimson tights so that blood from the occasional bullet graze wouldn't show. At the end of 1892 they all went off to Paris where her husband, who liked to be known as 'Le Roi des Cowboys', organized a series of 'challenges' against the bicycle-mad French. Horse versus bike was the idea – or, rather, their top racing cyclists taking him on, mounted, at venues like the Casino de Paris and Le Levallois-Peret Trotting Club. At the former he beat the great champion Terron (who had recently ridden from St Petersburg in record time) to take a FF10,000 prize while standing astride *two* charging thoroughbreds. 'Vive Cody!' cried the crowd. Then, in 1894, accompanied by a string of yearlings and five grooms (one 'a full-blooded Red Indian'), he travelled to Germany and vanquished fifty of their ace cyclists with ease.

Back in Paris he was challenged by the famous walker Galleaux to an endurance test – a fifty-hour race, the man on foot against Cody on his stallion Bergamo. The French public grew so enthralled the French press produced special six-hourly editions and naturally, in their final one, reported the victory of a yawning Cody and his sleep-deprived mount. The next contest, between Cody and a two-man team consisting of a French cavalry major and another full-blooded Indian (named Humpa), was a three-hour dash in which each side was allocated twenty horses. Humpa and the major, every few moments, relieved each other with fresh mounts while Cody, galloping furiously, just kept vaulting from saddle to saddle, never touched the ground; unsurprisingly, he came home first.

In 1895 Cody, visiting Rome, claimed to have revived chariot-racing. 'In a race of six teams, each with four horses abreast, I managed to get first place, and my son Vivian – then only ten years old – was third. In Malta my son Leon won a big race when he was eleven, riding bareback on an English mare against the Maltese professional jockeys on their native ponies.' In Malta Cody did some professional wrestling, often taking on – and beating – two opponents at a time, in Tunis had 'no end of excitement and grand sport, in

Cody, Lela and the three boys in a publicity shot
for one of his Western shows.

racing with horses, chariots and cyclists'. They returned to England
to see the celebrations for Queen Victoria's Diamond Jubilee.

The boys, who now affected Cody's long, tangled hair, hand-
tooled boots, buckskins and sombrero, were becoming variety stars
in their own right. Leon, swinging upside down from a trapeze, shot
moving glass targets at the Camberwell Baths while Vivian, being
dangled by his ankles, displayed his marksmanship at London's
Alhambra Music Hall. Cody's own newest trick was shooting ciga-
rettes from Lela's lips as, both mounted, they rode flat out in different
directions. Then, tiring of music hall, he moved into theatre. His first
melodrama, *The Klondyke Nugget*, premiered at St George's Theatre,
Walsall in 1896, featured blazing bridges, leaping horses, Indian
war parties, ferocious duels and a beautiful heroine – played by Lela
– being snatched from the jaws of death seconds before the final
curtain. It was judged a huge success, and made him a great deal of

money. So did its four successors – all plays from, ostensibly, Cody's own pen. His business card said, 'Samuel Franklin Cody, Dramatist of the Theatre Royal, Stratford E.'

*Dramatist*? Cody could neither read nor write.

Browsing in a London toy shop one day he bought, on impulse, a kite for the boys (did he also bring them stories of a Chinese cook?) and saw that if he made certain structural changes, it would go higher. Determined to go higher still, he built an entirely new kite, then a whole flock of kites, each bigger and better than the last. He flew them during every spare minute: between matinees and evening performances, in a Liverpool park, on the beach at Blackpool, and from the roof of the Metropole, Glasgow – where, down in the streets, pedestrians gazing distractedly upwards kept walking into each other as cyclists got entangled and people on roller skates painfully collided. Yet Cody, oblivious to the cries and clatter below, carried on; it was in Glasgow he told a well-wisher, 'It is my intention, sir, to build a kite that will carry a man.'

At Alexandra Palace, sharing a shed with a French balloonist named Gaudron – who, in 1908, would fly non-stop to Novo Alexandrovsk in Russia – he began work. Made from American hickory, bamboo, silk, steel tubing and piano wire, his prototype took him to a height of 80 feet before spinning and crashing into a tree. Its successor was a multi-tiered giant connected by a stout cable to a heavy winch. It had a basket seat which, 'for observation purposes', went whizzing up and down the cable like a fairground ride, while bridle lines connected to the trailing and leading edges gave him some element of control. And it worked. The world's first successful man-lifting kite, its black tiers resembling a line of baby bats in pursuit of their mother, was cheered by those who witnessed its maiden flight at Plumstead. The British government, however – whom Cody hoped to impress – took absolutely no notice.

Yet the navy invited him to Portsmouth where, after Leon had gone up 800 feet and photographed the warships, several officers went up too and commented on its potential for gunnery. Cody,

Cody, with man-carrying kite.

making an ascent from the stern of a cruiser, almost drowned when his kite plunged into the sea yet, unfazed, built a kite-powered canoe and, one evening, set off from France intending to cross the Channel. Midway over he became becalmed, hung a lamp from his mast (big ships kept thundering by) and drifted with the tides till dawn. Then the wind freshened, so that towed along by his tugging, bellying fifteen-foot kite he arrived in Dover in good time for breakfast. Whitehall still paid him no heed.

In 1901 he patented his man-lifter, and a wing-warping arrangement that assisted lateral control. (The Wright Brothers would come up with precisely the same idea in 1903.) That same year, having reached 300 feet – and survived another crash landing – he claimed to have made 'The highest ascent known to man without the aid of gas', and offered his 'Aroplaine [sic] or War-Kite' to the War Office. Their dusty reply was tempered by a welcome letter from the Royal

Cody shows off the canoe in which he crossed the English Channel.

Meteorological Society who, after he had managed to send some of their instruments higher than they'd ever gone before, awarded him a Fellowship. (His interest in meteorology, he claimed, was kindled by the Sioux Indians. As a young man he was shown the Sioux's Cave of the Winds, a grotto with an unusually narrow entrance. The tribal forecasters explained that if they could hear the whistle of air being forced in it meant rising atmospheric pressure and fair weather, but if the air was being sucked out – usually with an audible *whooosh* – it meant pressure was falling and storms would follow.) This story is probably no truer than the one recounted in *Pearson's Magazine*. 'My earliest recollections of kite flying', he told them, 'date from 1864, when I was three years old, when my father, returning [from] the Civil War, gave me my first sight of a kite in the air, after a thieving raid on his cattle by Redskins.'

Midway through 1903 he competed in the International Kite Trials on Worthing Down in Sussex (he came second, and Leon third), later, at the Stanhoe Hotel, Worthing, he and Colonel John

Capper were elected members of the Aeronautical Society of Great Britain. It was his first meeting with Capper, who was to become a figure of immense importance to him.

Lela, her feet planted in flower baskets, made a remarkable number of ascents. So did her boys. Leon, in fact, broke the world record for altitude in a man-lifting kite.

Cody intended that his giant kites be used to pull sledges in the Arctic regions, and flat-bottomed boats up fast-rushing rivers. To prove his point he had himself and the boys towed up a stream and onto a lake during a duck-hunting holiday in Ireland, but their arrival at the lake was so silent the ducks never heard them coming. The boys, forbidden to shoot them on the water, had to whoop and holler so they could be shot on the wing.

One day Cody noticed a small item tucked away in his morning paper. It announced that Messrs. Wilbur and Orville Wright of Ohio had successfully launched a heavier-than-air machine at Kitty Hawk, North Carolina. Advances in the aeronautical properties of box kites, the report concluded, had been incorporated in the construction of their airship. Cody, who had never made any secret of his progress, read it thoughtfully.

Early in 1904 Whitehall, at last offering a measure of support, invited him to join the Balloon Factory. At Farnborough he could hear the drums and bugles from Aldershot's North Camp, and sporadic gunfire from its ranges; there was space galore, the air was sweet and folks welcoming. It suited him just fine.

I'd never visited Farnborough before and, after my meeting with Jean Roberts, decided I should start getting to know the place; when I asked a cab driver for the town centre, however, he dropped me off at a semi-derelict shopping mall containing dozens of boarded-up businesses. Among those still trading was Lucky Star Tattoos, its door bearing a price list for the insertion of studs and rings ('Nipple £20 pair, Nose £10') and some stern banning orders: 'No children. No drugs or drinks. No attitudes'. At the back, bathed in bright light, a plump, dark-haired, white-skinned woman in a black bra was having a seahorse etched on her shoulder. I went in. The walls were papered with an assortment of blood-curdling gothic designs and, as I pondered them, a sniffling teenage girl in knee-high boots appeared and said, 'Yes?'

I was about to ask, in jest, if they did tattoos of old planes (though not with old needles) but she gave me such an odd, narrow-eyed stare I said, 'Just looking, thanks,' and returned to the mall. There a palpable air of gloom may have had something to do with a chill wind, and rain-bearing indigo clouds swirling so low they almost brushed the rooftops. More likely, though, it was the empty display windows and 'Acquired for major town centre development' notices, a strange air of abandonment. A worrying proportion of the passers-by were pensioners and doleful middle-aged folk evidently dressed by the charity shops; surely, I thought, there must be another part of town where the younger, smarter, better-off citizenry spent their money, but a frail-looking old man I chatted to told me they all went to Guildford.

'Or Farnham,' said his wife.

The man said, 'All this belongs to some Arab.'

'Nobody was consulted,' said his wife.

'Nobody even *saw* the bastard. Now he'll just fly in to collect the money.'

'More likely send one of his wives,' said his wife.

A lady at the functional-looking Farnborough Library said, 'We have no Cody section as such. But there are over five thousand books in the Aviation Collection, and I'm sure he'll feature in some.'

Upstairs I found a minstrels' gallery fairly bursting with stuff about flying – many from the archives of the Royal Aircraft Establishment. I riffled through *Who's Who in Aviation History* (Cody got a mention), *The Daily Telegraph Book of Airman's Obituaries* (not there though), five volumes of *The Air Annual of the British Empire*, and *Air Travel: How Safe Is It?*. At the next desk a beautiful young Chinese woman sat at one of the computers installed for public use. Idly noting the way her fingers raced across the keyboard I realized, with a start, that she was silently weeping. I asked, 'Are you all right?' then, as the tears coursed down her cheeks, knew how dumb that sounded. I said, 'Look, if there's anything I can do . . . ' but she gave an angry little shake of the head. I shrugged and went looking for the station. It was time to go home.

On Salisbury Plain, assisted by Leon and Vivian and a platoon of Royal Engineers, Cody built and demonstrated kites for the army. Clad in his usual buckskins and spurs, with a sombrero planted on his hippy-length hair, he supervised the ascents from the back of a huge white horse named Vichy. Periodically, to entertain the soldiers, he would fast-draw his pearl-handled Colt .45s and, despite constant threats of prosecution for defacing the coin of the realm, blast sixpences from the sky. The threats came from certain senior officers who detested Cody – a particular gripe being that the public, mixing him up with Buffalo Bill, kept calling him Colonel.

Cody, the perennial showman, supervises soldiers on Salisbury Plain.

The ordinary soldiers, however, adored the eccentric, democratically minded Yank who worked them hard yet always treated them as equals; the Cody Kite Duty became a sought-after job.

In 1906 he was appointed Chief Kiting Instructor to the British Army and allocated a workshop over the turnstiles at Crystal Palace station. (Both workshop and turnstiles have long gone.) His job was to provide the Royal Engineers with vehicles for recon-naissance. Since 1879 they'd been using balloons which couldn't function in winds exceeding 20 mph; Cody's kites, though, took even a 50 mph Force 10 in their stride. Such a storm – toppling trees (and deflecting artillery shells) – must have provided a wild and terrifying ride, but the men sent aloft seemed not to mind. In better weather heights of over 2,000 feet were routinely achieved, while a breathless Captain P. W. L. Broke-Smith once got to 3,000 feet, or almost two-thirds of a mile: a new world record for a kite-pilot.*

Next Cody fitted a kite with a rudder, front elevators and a propeller driven by a 15hp French Buchet engine. In 1907, with nobody aboard, it chugged over Farnborough Common for four and a half minutes. Cody, who had now made over two hundred kite ascents,

---

* Many of Cody's kites may still be seen in the climate-controlled archives of Seattle's Drachen Foundation. (Drachen means 'kite' – also 'dragon' – in German.) An educational establishment 'devoted to the increase and dissem-ination of knowledge about kites worldwide', it undertakes preservation work such as the refurbishment of Harvard University's nineteenth-century kite shed, also scientific research into kite-related matters like, for example, certain high-flying American bats a kite might tangle with. Its archives contain historic kites from Japan, Cambodia, India, Laos and China, Thai royal kites, Korean fighting kites, traditional leaf kites from Oceania – possibly the world's first kites – and even a tetrahedral kite cell made by Alexander Graham Bell. But the jewel in its crown, its most important single acquisition, is the Cody Collection with a catalogue running to thirty-nine single-spaced pages. Acquired at Sotheby's in 1997 in the face of stiff opposition from various museums and col-lectors around the world, it comprises not just his beautiful old kites but also his exquisitely drawn kite plans, along with glass-plate negatives, posters, press clippings and business records.

felt the time had come to move on. Powered flight was the way forward. First, though, one had to learn how to master a heavier-than-air machine. He built an enormous glider – essentially just a big wing, launched from a cable and looking, with its subtle curves, elegant angles and flexing sweeps, like a piece of descriptive geometry worked in bamboo. Cody, slung underneath on a rope hammock, flew it many times, and so did the Cody boys. Vivian, lightly built, and with those trick-shooter's quick reflexes, was turning into an excellent pilot, intuitive and, of course, self-taught. One morning he took it up, stalled – it is thought – then flew vertically into the ground from a height of 800 feet. He nearly died and Cody, badly shaken, decided that his future machines required coercion, propulsion, a bit of beefing up. They would need, in short, engines.

He formed a close friendship with Sir Hiram Maxim; the two, of course, had much in common, and Albert Peter Thurston, an engineer who had worked with both, later wrote, 'It was delightful to see the action and reaction of Cody on Maxim and Maxim on Cody. Cody's gadgets were extremely simple and ingenious but the workmanship was poor; Maxim however was the finest craftsman I have ever met. I remember Cody showing Maxim a gadget for deflecting the running rope of a kite. Maxim approved the design but pooh-poohed the workmanship and said he would make one for him. Maxim took Cody's model straight to Norwood, worked nearly all night and next morning . . . I heard Maxim say to Cody: "There you are, that's the way to make it." '

When Colonel John Capper, with whom Cody had been inducted into the Aeronautical Society at Worthing, became Commandant of the Balloon Factory, one of his priorities was the completion of *Nulli Secundus*, Britain's first dirigible, or steerable, balloon. Begun years earlier by Templer (after a meeting with Alberto Santos-Dumont, who had flown an airship around the Eiffel Tower) then abandoned due to cost, it was a hundred and twenty-two feet long and made of

laminations of goldbeater's skin – the tough, lightweight material created from the intestines of animals. (Over two hundred thousand
cattle had been slaughtered for *Nulli Secundus* which looked, and
doubtless tasted, like a giant sausage.) Capper needed someone to
design the control surfaces, find an engine and get the thing airborne.
He knew just the man.

Cody travelled to France for the engine, a 50 hp Antoinette
designed by a M. Leon Levavasseur. He and Cody hit it off at once.
Levavasseur, a French monarchist and artist turned engineer, insisted
he stay as his house guest. Funny, subtle, subversive, endlessly inven-
tive – and with the same hulking physical presence – Levavasseur
so charmed Cody that he turned into a French monarchist too
(unaware that, one day, the French Empress Eugénie would arrive,
newly widowed, in Farnborough, and become a neighbour and one
of the great loves of his life). Levavasseur, who understood the kind
of people running the British government, sent him to a barber to
get his hair cut and his bushy black beard trimmed. Cody arrived
home with the French goatee and lavender-waxed moustaches he
would wear for the rest of his days – and a gleaming state-of the-art
automatic fuel-injected, eight-cylinder power plant that had, actually, been intended for a motor boat.

After its installation in *Nulli Secundus* a pair of counter-rotating
propellers had to be devised. He bolted spade-like blades to T-shaped
tubes then mounted them on outriggers. Chain drives with sprockets
– the chains crossed to ensure the propellers revolved in opposite
directions – worked through tubes filled with tallow. On 10 September 1907, the dirigible was taken, gingerly on the backs of
soldiers, to Farnborough's golf course for a tethered test flight. In
the course of other brief ascents minor adjustments were made to
the drive chains. (Cody also fitted an umbrella-shaped stabilizer
to prevent pitching.)

Each morning at dawn Capper would march outside, light his
pipe and ponder the behaviour of the smoke. Dead calm conditions
were what he wanted and, on 5 October, he got them. Capper, Cody,

*Nulli Secundus I* startles London in 1907.

and a ballooning instructor named Captain W. A. de C. King
scrambled aboard the open canoe-shaped gondola. Cody tended
the engine (he alone had the brute strength to get it started) while
King, reading the road map, called directions to Capper, who sat
at the helm. Moving at a comfortable sixteen knots they passed
over Chobham, Staines and Chiswick, followed the Thames from
Fulham to Lambeth, flew across Whitehall and the War Office
(where the Army Council stood on the roof, elatedly waving their

handkerchiefs), circled Buckingham Palace then – as traffic halted and a sea of faces gazed up – arrived over St Paul's.

Turning south-west for home, however, Capper felt a head-wind pushing them north-east into Essex. A landing on Clapham Common was out of the question due to crowds converging, so they made for the Crystal Palace. There Cody, throwing a line and shouting instructions to passers-by, got them safely down. 'England Mistress of the Air!' crowed a jubilant press – a little prematurely, as it turned out. Next day it rained, and the goldbeater's skin was found to absorb moisture at such an astounding rate that *Nulli Secundus* rapidly became a sodden hulk. Before it imploded under its own weight a wind sprang up and the thing began to thrash around. A Sergeant Ramsey ripped it open with his jackknife and, to the horror of the War Office, released 55,000 cubic feet of priceless hydrogen into the atmosphere.

What interested Cody, however, were thousands of tiny spiders that had attached themselves to the dirigible. Each had spun an eighteen-inch-long diaphanous wing capable of catching a thermal, lifting the spider to a good height and carrying it a significant distance. They had been hitching lifts on *Nulli Secundus*, and now he found himself pondering the buoyancy of gossamer.

In 1904 Capper and his wife had visited the Wright Brothers in Dayton, Ohio. 'Both these gentlemen', he wrote, 'impressed me favourably . . . they are . . . well-educated men and capable mechanics.' (Mrs Capper certainly impressed Wilbur, who found her 'an unusually bright woman.') Capper, home again, badgered the War Office to buy a Wright Flyer and persuade the Americans to continue their work at Farnborough. But their terms – $100,000 for a single machine (though it would entitle the War Office to build more) plus a further $100,000 for access to their 'scientific knowledge and discoveries' – were received with incredulity in Whitehall; Capper, who had his own American prodigy, finally realized he didn't need them and, towards the end of 1907, the War Office's Director of Fortifications agreed. Cody was to begin work on a project to be known as British Army Aeroplane No. 1. Capper, head still reeling from the stupendous figures Orville and Wilbur had been bandying about, pointed out it would need money.

The Director of Fortifications gave him £50.

Cody bagged a corner of the Balloon Factory and, helped by Vivian, Leon, and Edward Leroy, an actor pal from his Wild West shows, laid out the machine. Its upper and lower wings were cross-braced with piano wire (calculations for the Euler strut and bridge truss were made by Colonel Capper, using his engineering training – which may not have included the Theorem of Three Moments, a sophisticated means of assessing how spars respond to stress), and it had a forward elevator to control pitching and a single rudder at the rear. The wings were surfaced in brown Holland silk by a pretty Farnborough milliner named Miss Eva Copeland (assisted by a Mrs Hannon, a Miss Witticase and a Miss Bromfield; there is a photograph of them at work wearing long skirts with aprons or pinafores, and warm woolly tops; that cavernous building was unheated). The War Office, after keeping him on tenterhooks for months, agreed to lend him *Nulli Secundus*'s Antoinette engine. Tethering the finished machine to a tree* with a spring balance, he would then open the throttle wide and weigh the power pound for pound, measure it in avoirdupois.

Orville Wright spent a weekend at Colonel Capper's home, yarning and relaxing. When, strolling in the garden, he admired Mrs

---

* For years Cody's tree, heavily marked by rope burns and oil, was preserved at Farnborough behind a stout iron fence. A tablet read, 'Col. S.F. Cody picketed his aeroplane to this tree and from near this spot on 16th May, 1908, [the date famously wrong] made the first successful officially recorded flight in Great Britain.' It attained, in aeronautical terms, the status of the one true cross; as it aged the faithful helped themselves to relics (the Royal Aeronautical Society has one), when it died an aluminium alloy replica – also known as Cody's Tree – was made by apprentices of the Royal Aircraft Establishment and erected in its place.

The elegant new Cody II, which crashed on one of its first outings.
But then, rebuilt, it went on to win the Michelin Cup (Cody, after almost
five hours aloft, returned to earth with his beard and clothing
coated in ice).

Capper's climbing roses, she had one potted and sent to his sister
Katharine, back in Dayton, Ohio.

Early on 30 September 1908, according to Penrose, a team of
horses towed Army Aeroplane No. 1 to a quarter-mile stretch cleared
by the Royal Engineers for their morning gallops. Sappers clung to
his wingtips as Cody gunned the engine then, when he waved,
stepped smartly back; his machine, lurching across the turf, seemed
to rise a few inches. Though it was immediately established that, for
seventy-eight yards, there were no wheel tracks in the dew, Cody
chose to discount it. 'It was only a jump,' he said.

'Only a jump,' said *The Times* next day.

On 16 October he tried again. 'The machine', reported the
*Morning Post*, 'was set full speed up the slopes of Farnborough Hill,
from the top of which it took a rise into the air travelling westerly.'
For the first time he was properly airborne, and at a height of about
20 feet and a speed of 30 mph avoided a stand of trees. But other,

taller trees loomed ahead, and the violence of the manoeuvre needed to avoid them – 'I turned the rudder,' Cody recalled, 'and turned it rather sharp' – caused him to bank so steeply 'the machine spun round and struck the ground that way on, and the framework was considerably wrecked.' Though cut and bruised he scrambled clear, grinning broadly, having been aloft for twenty-seven seconds and covered a distance of 1,390 feet.

Cody was contrite about his smashed machine, yet exhilarated too. *He had just become the first person in Britain to fly an aeroplane.* And though Capper had witnessed that historic moment, he seemed more concerned with the crash report he must now send to the Director of Fortifications. That evening, teeth gritted, he wrote, 'The damage done to the machine is: left wing, good deal broken up, silk some-what torn; head rudder stays damaged, right wing slightly damaged; wheels buckled; engine fly-wheel broken.' Gamely he added, 'I do not propose to abandon trials with this machine.'

He had another equally difficult letter to write. J. W. Dunne, working at Blair Atholl on his inherently stable biplane, had long dreamed of becoming Britain's first aeronaut (since, latterly, his dreams had become aspiring and non-prophetic, Capper was able to dream with him). He would be devastated by tomorrow's press reports, so words of comfort and encouragement must be in the post tonight; Capper still believed the future of British aviation lay with his clever, polished, upper-class friend and fellow veteran – and certainly not with the cocky American buffoon at the Balloon Factory.

In Germany an airship four times as large as *Nulli Secundus* exploded in a storm. The German government attributed its end to an Act of God. But, much impressed by its range, power and military poten-tial, they awarded its designer, Count Zeppelin, the Order of the Black Eagle and asked for eight more.

Uppermost in Capper's mind was the undeniable fact that Cody couldn't fly; and – but for trial and error – had no means of learning. Cody, though, remained unperturbed. When he bought his first car he raced it, from Central London to Farnborough, without having had a driving lesson in his life. His instincts, he reckoned, were sound.

When, eventually, the repairs to his machine were complete, Capper made Cody taxi around to get the feel of the controls. One day, during a moment's inattention, he knocked a mounted policeman off his horse.

## Chapter Nine

# Straighten Up and Fly Right!

Cody about to leave for Manchester. Shortly afterwards
he flew into a hedge.

Early in 1909 at least forty Britons were building aeroplanes, while
potential customers like Charles Rolls and John Moore-Brabazon,
who dreamed of winning records and prize money, lurked in the
background, watching them.

And they all watched Cody.

He was coming along, in his fashion. He crashed on 9 January,
again on 21 January. Colonel Capper wrote a further brave letter,

anticipating 'a good many smashes before Mr. Cody has learned to manipulate the controls.' He concluded wearily, 'Though I make suggestions at times it rests entirely with him whether he follows them or not.'

After each incident, however, back in the Balloon Factory repair shop, he remembered what he had done wrong. Journalists mocked him for his 'mowing machine', for his agricultural habit of digging up fields and knocking down trees. Some resented a loud-mouthed foreign upstart spending British taxpayers' money on a plane he couldn't fly, and tried making him a laughing stock (he never forgot that, and when journalists summarily bundled out of Blair Atholl described the Duke's private army as 'savages', he was enormously cheered). But the British public loved Cody's exuberance and glamour, and the sheer romance of what he was trying to do. (Soar above the rooftops!) Today he would be an A-list celebrity, endlessly written up in the gossip columns, profiled in the broadsheets, popping up on television offering opinions on every subject under the sun and – being an utterly shameless self-publicist – relishing every moment. And, goodness, how he would have relished the way Britain, after forgetting about him for almost a century, once again began showing respect.

The Cody Astronomical Society was founded in 1998. Its finest hour occurred in June 2004 when people from miles around queued to witness the Transit of Venus through its six-inch Carl Zeiss Coudé refractor. The Society holds regular meetings, all posted on its notice board (the cancellation of one – 'Lilian Hobbs can't give her lunar lecture' – made me smile) and training sessions for novices who are 'shown around the sky.'

The Cody Farnborough Amateur Operatic Society performed its first show, *The Golden Amulet*, in 1921. Then it was known as the Royal Aircraft Establishment Dramatic and Operatic Society and many of its early stars were Balloon Factory veterans. Of the hundred productions it has mounted perhaps the most unusual was at the Theatre Royal, Aldershot, during World War Two, when it shared the bill with Phyllis Dixey, a well-known striptease artiste. The troops flooding in to see her found themselves having first to sit through a Gilbert & Sullivan operetta – which, quite unexpectedly, became the talk of the town one night when Pooh-Bah, kneeling before the Mikado, said, 'Your Majesty, I have to announce that the sirens have just gone off.' The company got a standing ovation and not a soul left the premises.

Today it performs two shows annually at Aldershot's Prince's Hall. (Recent successes include an Australian adaptation of *HMS Pinafore* featuring the Absolutely Fabulettes, a female comedy trio.) The Society holds regular beer and skittles evenings at the Jolly Miller pub, North Warnborough.

I had Googled the Samuel Cody School in Lynchford Road, Farnborough (accessing its site via a little Cody biplane which, when clicked, dipped its wings prettily), so I knew it catered for children with special educational needs. Some had autistic spectrum disorder, a learning difficulty ranging from speech defects (a few never speak at all) to a chronic fear of relationships, others various behavioural and emotional problems. Its syllabus bore a striking resemblance to that at Brompton Hall, and walking one autumnal afternoon down a driveway piled with slippery, russet-gold leaves (a stray sunbeam made them look new-minted) I was touched by the fact that England's two greatest aviation pioneers were commemorated by establishments for helping troubled children.

The school was a two-storeyed, flat-roofed brick structure overlooking a neatly mown playing field. An arrow pointed visitors to a rear door and, making my way there, I saw a small girl watching me from an open upstairs window. She had blonde pigtails and glasses, and a faintly distracted air. 'Hullo!' I called, but she didn't respond. The visitors' door was electronically locked, though off the lobby an office had been left with its light on and door open. I sounded the buzzer for several minutes then, turning away, saw the child still standing at the window. 'Is there a teacher in your classroom?'

I asked. She gave no sign of having heard. I said, 'Do you know who Samuel Cody was?' Now she smiled faintly. 'The king, the king, the kingalingaling!' she declaimed then, turning away, closed the window behind her.

The Cody Society was founded in September 1993 by a group of enthusiasts who set out their aims on a website. These were '1. To honour the name of Colonel S F Cody in any manner deemed fit and proper. 2. To hold lectures, meeting [sic] and discussions from time to time', also to plan a 'commemorative Service of Remembrance at the site of the CODY Tree', meaning the aluminium replica in front of Farnborough's Black Sheds. It was there, on 24 October 1995, that His Royal Highness Prince Michael of Kent KCVO, along with the United States Assistant Air Attaché, US Air Force Colonel Timothy Moore (plus Mrs Moore) and various local dignitaries were welcomed by society members as music was provided by the Band of the Royal Logistics Corps. I contacted the society's founder, a strong-voiced ninety-two-year-old who told me he had cancer.

'I'm very sorry to hear that,' I responded. 'But could I perhaps visit you to talk a bit about Cody? Just for a few minutes?'

'Out of the question,' he said. 'What I've got is terminal. I'm dying, you know.'

Eventually Cody discovered the secret of setting down in one piece: 'You must imitate the squatting posture of a crow.' That directive went rippling out as far as Essex. Alliott Verdon Roe told his friends, 'Cody says you should land like a crow!' and immediately applied the principle to his own turbulent touchdowns.

Turns, though, still bothered Cody. 'I am not confident,' he told the *Morning Post*, 'that I have quite discovered the amount of banking this machine requires in order to keep its balance in turning movements.' (The *Morning Post* was one of only two or three papers – *The Times* was another – he trusted. He kept all his press cuttings, and appeared to know them by heart; when someone who had written unfavourably about him appeared, Cody, brandishing the evidence, would order him off the premises.)

By April he could turn well enough to do a complete circuit of Laffan's Plain – the longest flight, as it happened, ever made in Britain. The Prince of Wales, in residence at Farnborough's Government House (today just across the road from the Wavell-Cody Community Campus), rode over to offer his congratulations. Cody the ardent royalist almost received him on bended knee.

The future George V said, 'I am very sorry I missed your flight. I should like to have seen that.'

'Then perhaps, sire,' said Cody, 'you will allow me to do it again.' Clambering back aboard, ever the showman, he delivered a low-level Royal Command spectacular that made the Prince's ears ring and frightened his horse. On landing, however, be scraped a grassy bank and bent a wing. The prince took a shine to Cody, from that moment followed his progress with interest and liked to drop in, unannounced.

A few days later, without warning, they sacked him. A committee of military men and politicians – David Lloyd George and the War Minister R. B. Haldane among them – was appointed by the Prime Minister, Herbert Asquith, to put aviation on a formal footing, or at least give Britain equity with France, where top pilots enjoyed the kind of celebrity we now accord rock stars. Having examined all the evidence they decreed (just five years before the outbreak of World War One, and plainly influenced by War Office attitudes) that aeroplanes were 'useless' for military purposes; from now on the nation's defence would rely on airships. The first thing needed would be a headquarters, and the Balloon Factory seemed an obvious place. Though a correspondent from the *Aero* found Laffan's Plain 'abominable', its surface 'bumpy and cut up by gullies, most of it on a sideways slant' with a surfeit of trees (Cody had endured this rustic cow pasture without comment or complaint, and suddenly, the old jokes about his piloting abilities seemed less funny), the committee, perversely, found in its favour. Back in Whitehall plans were announced to turn Laffan's Plain, Cove Common and Long Valley into a single entity, a proper base – for dirigibles.

It was R. B. Haldane who broke the news to Cody. He had an excuse ready: his contract with the War Office had expired and would not be renewed. Cody, however, could keep his machine – though the engine must be returned – and if he wished to continue flying they were prepared to let him use Laffan's Plain. He then turned to a distraught Colonel Capper and accused him of spending the bulk of the Balloon Factory's £2,500 budget on J. W. Dunne and his quest for inherent stability.

Cody knew that Capper had always favoured Dunne, yet his loyalty to the colonel never wavered. But he remained a can-do Yankee optimist and, in May 1909, prepared to go it alone. Under the pines at the Fleet Road end of Laffan's Plain he erected a shed large enough to hold his battered aeroplane, then put a workshop at the back, and an army tent beside it. Cody lived with Lela and the boys at Pinehurst on Mytchett Road but knew that, some nights, he'd be working so late it wouldn't make sense to go home.

In the workshop his Army Aeroplane underwent some radical changes. The wings and chassis were entirely rebuilt, and while he retained a steering wheel to turn the rudder, he now had a big hickory skid set under the tail, and in place of the bucket seat behind the engine, a tractor seat in front (encouraging the press to wheel out all their old Ploughman Cody wisecracks). The new engine, which took all his savings, was French. But his new, markedly lighter, plane flew wonderfully well and, routinely making trips of four miles, caused embarrassment at the War Office; there, no doubt, some of those responsible for the government's anti-aircraft line composed (just in case) elegant little memos deftly redefining their positions.

People dubbed it 'the Cathedral'. Various reasons have been put forward for this, the most plausible being that it originated with a Prof. F. W. Lanchester, who was impressed by its size and the graceful katahedral angle, or downward sweep, of its wings.

Cody fitted a second tractor seat and began giving rides. The first went to Colonel John Capper (making him Britain's premier air passenger), the second to Mrs Sam Cody – unsmiling in a heavy coat and bonnet, she was flown to the end of Laffan's Plain and back carefully as a box of eggs. Over the course of the next few weeks he took the boys up, along with various friends, some of Lela's relatives and, possibly, Sir Thomas Lipton and Mr Gordon Selfridge. Cody now attracted such crowds to Laffan's Plain that Farnborough's Military Police were needed to control them.

One particular visitor was an old lady who had recently built an abbey, St Michael's, in the French Gothic style, at Farnborough, and bought a stately home, Farnborough Court, nearby. The Empress Eugénie, Spanish-born widow of Napoleon III – France's last emperor – was in her eighties when she became Cody's neighbour. He, being a fully signed-up French royalist, was besotted by her presence. There is a picture of her, all in black, bony, animated and immensely striking, inspecting his aeroplane in 1909; he, head inclined deferentially, almost bursts with pride and excitement.

Theirs was an intense friendship. She may, possibly, have helped him with his reading and writing, but suggestions that there was a romantic attachment are nonsense; he was far too awestruck, she far too old. In the 1938 movie *Suez* a much younger Eugénie (played by Loretta Young) was portrayed as the lover of the canal builder, Ferdinand de Lesseps. (Whether that happened in real life is unclear; Eugénie was, in fact, bisexual, having had affairs with the wife of Austria's Ambassador to France, and the English composer Dame Ethel Smyth – a promiscuous lesbian who seduced Virginia Woolf, and whose best-known work was an opera called *The Wreckers*.) Eugénie created St Michael's Abbey as a mausoleum for her husband and son, the Prince Imperial, who, aged twenty-three, died fighting for the British in the Zulu Wars; seventeen assegai wounds to the front of his body demonstrated the courage with

Cody conducts the Empress Eugenie around one of his machines.
An ardent French monarchist, he was overjoyed when
she settled in Farnborough.

which he met his end, and he was entombed here in the family
crypt, wearing British uniform. Today his mother lies adjacent to
him, dressed as a nun.*

I'd always believed her *basilique impériale* to be off-limits. Set high
over the town and shrouded by trees, it was guarded by large, locked
metal gates. Then I learned that at 3.30 p.m. each Saturday the gates

* The pair of them have given rise to a curious little tale. The first asteroid
to be named for a living person – rather than a figure from classical legend –
was dedicated to Eugénie in 1875 by the French academicians who discovered
it. Eugenia 45 is a big Main Belt, F-type asteroid, which means it is dark in
colour, with a carbonaceous composition. It has a diameter of 214 kilometres,
and an unusually low density indicating it may be loosely packed rubble rather
than a monolithic object. In 1998 astronomers in Hawaii were amazed to spot
a tiny moon – its diameter only 13 kilometres – in a near-perfect circular orbit
around it: the first asteroidal satellite ever sighted from earth. The astronomers,
inspired by Saint-Exupéry's classic fable (published in 1945), approached the

swung open and the monks offered free guided tours. So, presenting myself one fine afternoon in May, I was able to walk unimpeded up a long, steep driveway which, once I'd rounded a curve by the farm and apiary, ended in a leafy, sunlit and distinctly surprisingly corner of France ('France transplanted,' said Monsignor Ronald Knox, who was received into the Catholic Church here). The abbey, incorporating half a dozen different French architectural styles, included a lovely miniature version of the cathedral at Tours, with windows of Flemish bottle-glass.

A dozen visitors had gathered outside the cathedral. Dom Thomas Harper, the abbey's guestmaster, wore a Benedictine's black habit; a small, hollow-cheeked man of indeterminate age, he led us up its vaulted nave, sat us on its uncushioned pews and for an hour – voice barely audible, staring fixedly over our heads – described every feature of its grand, feature-filled interior. His was a silent order, not known for its charismatic public speaking, and only afterwards, as he took us outside to see snarling Gallic gargoyles and the monks' tiny graveyard, did he grow animated. 'When pilots are practising stunts for the air show,' he said, 'they often do it right above this church. It's a focal point for them. I love to stand and watch.'

I said, 'Do you know if Sam Cody ever came here?'

'I'm sure he did. He and the empress were good friends.'

---

agency allocating Minor Planet Names and asked that it be called '"Petit Prince" in honour of Napoléon Eugène Louis Jean Joseph Bonaparte (1856–1879), Prince Imperial of France, only child of Empress Eugenie, namesake of Primary Asteroid 45.' They pointed out similarities between the Prince Imperial and Saint-Exupéry's boy mystic. Both were young, small and fearless. Both journeyed to Africa – the Little Prince from Asteroid B612, the Prince Imperial from Chislehurst, Kent – and died there: one bitten by a snake, the other speared by Zulus. It was, argued the astronomers, 'cosmically fitting' that Eugénie be reunited with her son 'in the heavens' since the baby moon designated S451 had probably come from Eugénie's asteroid anyway, caused by a stray comet, or 'collision of an impactor with the primary'. The Minor Planet Names Agency agreed, and Eugenia's moon is now known as '(45) Eugenia 1 Petit-Prince'.

Finally he drew a key from his pocket and opened the door of the crypt.

It was what we had come for.

The light was dim. On the floor an inlaid marble star duplicated the one surrounding Bonaparte's tomb at Les Invalides. The massive granite sarcophagi of Napoleon III and the Prince Imperial were placed in transepts, while Eugénie's sat on a marble shelf above the altar. Dom Thomas pointed to a graceful wooden staircase winding down through the gloom. 'She had her own key, her own door, her own stairs – that's a Destailleur *escalier*; friends would have entered with her there. Eugénie was a sad figure, a widow for forty-seven years, a grieving mother for forty-one. But her funeral, in 1920, was a great affair, a private train brought her body to Farnborough, soldiers lined the route from the station, George V and Queen Mary were here with all the crowned heads of Europe . . . '

One of our group, a wiry, earnest young New Yorker, interrupted to say that as fifty Zulu warriors charged towards the Prince Imperial – and his British cavalry escort vanished like the morning dew – he had, inexplicably, dismounted. His wife, who was short and broad-shouldered, beamed. 'Art knows all about this.' Art continued, 'But why? Did he want to talk to those guys? Did he have some Zulu jokes for them? All we know is he then changed his mind, put a foot in a stirrup, and . . .'

'His saddle broke.' This came from an amused-looking Asian in a blue shell suit.

Art's wife spoke tartly. 'Actually, sir, it *was the stitching on his surcingle* that broke.'

'It's often alleged,' Art continued, 'that his mother was too mean to buy him a new saddle. Not true. He'd chosen to use an old one of his father's, and it was just plain worn out. The Zulus said he fought like a lion.'

There was a photograph of him, there in the crypt. He looked like Errol Flynn.

'That may be,' said Dom Thomas, 'but they inflicted terrible injuries, stabbed him in the stomach, the chest, the head and face. They cut the boy to pieces.'

There was a moment's silence as we pondered the tomb in which the boy lay – a gift from Queen Victoria, who had adored him.

Art's wife said, 'There's a story about his pocket watch. It was taken by one of the Zulus who heard the ticking and thought it was a cage with a small animal inside. But he couldn't open it and, quite soon, the animal "died". Now it's in a museum somewhere.'

I asked Dom Thomas how the French felt about their last king lying in a church in Hampshire. He said, 'It's always been a political issue. Every once in a while there's pressure to return the Imperial Family, but our prior says no, and he's always supported by the Princess Napoleon in Paris. She also thinks they should be left undisturbed.'

Our tour ended at the abbey shop. It had its own website, and our shell-suited Asian friend produced a scribbled list of items he wanted. (They included Fast Lighting Incense Charcoals and a 'Nuns Having Fun' calendar.) Among the religious books on display were several that had been produced in-house: the abbey, Dom Thomas said, had its own printing press and bindery. I chose items produced by its free-range hens and long-range bees, and walking away past a beautiful albino peacock posing on a five-barred gate, had a sudden, cheering vision of Cody. He was bearing eggs and honey too – in his case gifts from the deposed Empress of France – as, stepping out jauntily, he headed home to Lela.

On 25 July 1909, shortly after five o'clock in the morning, a tiny aircraft came heavily to earth near Dover Castle. The pilot, weary and oil-smeared, gave his name as Blériot, Louis. Britain's aloof island status had ended forever, at the War Office there was profound shock. The role of the aeroplane in the defence of the realm might, after all, need to be reconsidered.

\*

On 8 September 1909, Cody flew, for an hour and six minutes, at a height of 800 feet – a new world altitude record. According to Broomfield thousands of excited people poured onto Laffan's Plain, cheering him 'to the echo'; he damaged a wing while trying to avoid them then, managing to get lost in the crowd, went and hid in his shed.

For this flight the Royal Aeronautical Society, at a gala dinner, presented him with its Silver Medal. One of the speakers was Colonel Capper, who savaged the British press. 'When Cody had a mishap it was really heart-breaking to see the ridicule with which his efforts were treated, and a man of less energy would have taken it to heart and chucked up the sponge.' Cody, accepting the medal, said, 'I, and the men of my family before me, have been very determined people, and, if I set myself a task, try hard to accomplish that task. I work at it, and I sleep and dream about it, until I have accomplished my end.'

On long winter evenings Lela and her sons tried to teach Cody, now one of the most famous men in England, how to read and write. He proved a fidgety, preoccupied student, easily distracted – but then, to their delight, began keeping a diary. 'Is this modern craze for arioul' went a typical entry 'a desiase that is catching lately? Wall I don't know realy wheather I am afflicted in this way but I assur you my air ship is never out of my deepest thought even dream of all sorts of aroplains and arioul flights amongst the clouds in fact if my dreams wer to come tru the moon would be so ashamed of itself it might sece to shine.'

How did it feel to be a pioneer making a maiden flight? Could one replicate the high drama of such a moment? Would the act of flying a small plane for the first time – over a part of Britain notorious for its Gust Fronts and Steam Fogs – offer any clues or insights? A decade ago, quite by chance, I stumbled on the answer.

The hundred and forty civil aerodromes that lead, like stepping stones, from Wick on the crown of Scotland to Land's End on England's westernmost extremity, had given me an idea. Forty one-hour lessons are deemed necessary for a private pilot's licence, so I'd devised a gypsy's curriculum which would enable me to start at one end of the country and, after visiting forty airfields promising some interesting characters (and colourful instructors), finish – hopefully in one piece – at the other. A novel way of seeing the British mainland, it could well yield a series of articles for the *Observer*, or a book – perhaps even both.

I arrived at Wick, for my first lesson, by train, on a fine late-August day, found a mellow little town where hot sunlight played on weathered stone and turned the slate-coloured North Sea a luscious blue. Half-deafened by unusually dissonant, rackety gulls I set off from its centre, and after a short walk came to Britain's most northerly mainland airport. In the tiny terminal a typewritten notice in a display case said that it occupied the site of Clayquoys Farm, which in 1939 had been cleared of walls, hedges and ditches and licensed as a landing ground. The remarkable thing about it, though, was that it appeared to be a farm still, with wheat stretching as far as the eye could see. It was ripe and ready for harvesting, and so tall that when a little BA Highlands and Islands turboprop touched down it was – but for the tip of its tail, which looked like a shark's fin skimming across a hazy golden sea – completely swallowed up.

I asked the girl at the snack bar where I could find Andrew Bruce. Advised that he worked in the old control tower I made my way there, spotted a man on a metal salient twenty feet off the ground. 'Up the ladder!' he cried, indicating that one reached the premises of Far North Flight Training via its fire escape.

Bruce, in his late thirties, was a stocky, combative-looking Scot with a profoundly restless manner. We stood for a moment on the salient looking out over the shimmering wheat. 'It's ready now,' he announced. 'In a day or two they'll get it in.' Then we entered an airy, spacious, sunlit room where a woman was talking into a radio. 'That's Moira,' he said, adding that he had learned to fly while still an Edinburgh schoolboy, at seventeen, after that – and this in tones of absolute indifference – becoming the world's youngest instructor.

The air, stupendously clear, allowed untrammelled views through two full quadrants of the compass. I could see the east coast, the west coast and, heather-swathed a few miles off, the mainland's most northerly point: Dunnet Head. From here we commanded the granite bows of the kingdom yet Bruce – fidgeting, chronically restless, maintaining a jumping-bean momentum – paid it no heed. And when a Cessna Citation executive jet whistled low overhead he was halfway down the fire escape before it was even on the ground. 'Norwegian air ambulance,' he yelled. 'Chartered by some Oslo millionaires over to shoot grouse.'

Moira, a handsome, good-natured, middle-aged woman, said, 'Refuelling is what we do now. Flying lessons are just a minor part of the operation.'

She took some calls – a farmer's daughter from the Borders, taught to fly by Andrew when they were both seventeen, would be dropping by that evening on her way to Papa Westray – and made us coffee. 'Yesterday evening,' she said, 'we had visitors also, a German couple in a Piper. They were making for Plockton but it was getting late so they unpacked a wee tent and slept, very soundly, by the wheat.'

Bruce returned, demanding food. Pizza was torn asunder and wolfed down as he led me to a small blue-and-white plane standing just yards from the tower. 'This is a Cessna 172, built in Rheims,' he said, 'and the first thing you do is a walk-around.' He pushed at the propeller, tugged at the exhaust, wagged the tail and rapped on the fuselage; it gave off a dull boom.

Moira appeared at the top of the fire escape. 'Doesn't Alex want his weather?'

'*Weather*?' He scowled at a tangle of high cirrus in a deep blue sky. 'What weather?'

We clambered aboard. Andrew, explaining the controls, paid particular attention to the split foot pedals: the lower halves worked the steering, the upper the brakes. He fished a key from his pocket and started the clamorous 145hp Continental six-cylinder air-cooled engine. We began moving forward. He said, 'Why are you heading for the car park? You'll have to pay. Use the pedals like I told you.' I stamped on a pedal. It was stiff, with a long travel, like the clutch of an old truck. The plane veered sharply left. I stamped on the other and it veered right. Working those pedals like a Wurlitzer star I took us sashaying past the air ambulance, now preparing to depart. Both pilots glanced up, smirking, and I toyed briefly with the idea of making a lunge at them. But we staggered on and, finally, reached a broad, deserted boulevard heading off through the corn. 'Brakes,' he said.

'Phew!' I sat back, drained. 'That was quite something.'

'Now open the throttle.'

I stared at him. 'I don't even know where the throttle is.' A dreadful thought took hold. 'Christ, I hope you're not expecting me . . . '

He indicated a small Bakelite knob at the base of the control panel. 'Just yank it towards you.'

'Andrew, if you think . . . '

'The Norwegian's waiting. You're holding up traffic.'

So I gave it a yank and everything vibrated wildly. The corn began sliding by, seconds later had become a rushing tawny blur with little vortices of heat distorting the runway ahead.

I glanced over at him, appalled. He sat with his feet tucked away, hands folded in his lap, eyes half closed, exuding a Buddha-like quietude.

'Keep it straight,' he said.

'Straight?' Unsure what to do with the pedals I did nothing. '*How?*'

'Like that. Aye, good. Now pull back.'

Past caring, I tugged at the control column and, all at once, we were rising serenely. Crikey! Scotland's own evergreen Arctic, an intricately ordered patchwork of grain fields and grouse moors, unfolded below as, up here, trumpets sounded.

It was my Cody moment.

I had steered a machine safely into the air on a fine day in one of the most beautiful landscapes I'd ever encountered; my sense of accomplishment would linger for weeks.

Bruce said, 'To make a turn you move the rudder pedal and control column at the same instant. Got that? OK, let's go right.'

As the Stacks and Reefs of Duncansby passed below I decided, fancifully, that the rich lavender light had been created by suspended heather particles catching the shine of distant Polar pack ice; bearing west we looked across to Scapa Flow, on Hoy, diminished by mist and distance, the Old Man seeming no higher than a fire hydrant. I was growing in confidence until, over the modest little Castle of Mey, abruptly he switched off the engine and we were tumbling giddily. 'This is a full stall,' he said. 'See what I'm doing?'

I saw nothing.

He concluded his terrifying manoeuvre. 'See what I did?'

I hadn't the remotest idea what he'd done but, frightened he'd do it again, said, 'Yes.'

After an hour spent banking through this calm and radiant air he talked me down – 'Close the throttle, I want you to stall onto the ground, pull back now, keep those wings level, it's *easy*' – just as he had talked me up.

We arrived back among the wheatfields with a tremendous klipspringer bound.

Back inside I saw Moira. 'Guess what! I took off and landed!'

She smiled. 'He always makes his first-timers do that.'

Yet my first take-off and landing were supposed to be key moments occurring midway through the narrative (and halfway down the country, say Oxfordshire; going solo would have come a bit later, around Dorset maybe), and now I saw little point in going on. I was, though, oddly content. On that glorious summer's afternoon I'd actually flown a small plane over Britain, and for a few moments – or until the nerves settled – gazed down with the same fearful intensity as the pioneers (wide-eyed behind their heavy-duty, bottle-glass goggles) once did. Now, ten years on, I can still recall the way I seemed to see every leaf and grass blade, a peculiarly vivid optical illusion probably induced by stress or a trick of the light. In these islands – when the weather permits – its aureate richness and diamond clarity can add depth and interest to a view, even seem to magnify it. The work of artists such as Constable, Gainsborough, and that great celebrant of light, J. M. W. Turner, bear eloquent testimony to that.

When, in October 1909, an aero meeting was held on Doncaster's racecourse, the Town Council wanted Cody so badly they offered him a £2,000 attendance fee. He accepted but, landing in atrocious weather, upended his machine, cut his face and lay pinned beneath it, bleeding. When rescuers pulled him clear he produced a rope, lassoed the tail skid and – bleeding still – hauled the machine upright. Next day, still at the racecourse, he became a British citizen. Though Lela and the children were Britons, and he had made his home – and reputation – in that country, his reasons, it must be said, had primarily to do with politics and prizes. As Britain found itself in the grip of aviation mania, races and competitions were being announced almost weekly, with serious money at stake: the *Daily Mail*, for example, promised £10,000 for the first London to Manchester flight; to win such events, however, you had to be British, and at the controls of a British-built machine – though when, early in 1910, they changed the rules it was, ironically a French ace, Louis Paulhan, who made it to Manchester first. (But only just; a humdrum record attempt became a thrilling race as Claude Grahame-White, the young man who had invented carrier landings at Crystal Palace, gave chase and almost caught him, their arrival triggering civic celebrations reminiscent of a United Cup Final win.) But, in 1909, the Britons-only policy still pertained; John Moore-Brabazon at Leysdown received £1,000 for flying the first circular mile in Britain when everyone knew that Cody, at Farnborough, had beaten him easily.

The judges, though, ruled that at the time he had been a citizen of the United States.

Mr R. A. H. Tovey, the Doncaster town clerk, agreed to offici- ate. On his arrival the American flag was hoisted and the band of the Yorkshire Dragoons played 'The Star-Spangled Banner'. Cody, facing a packed grandstand, swept off his cap and took the oath of allegiance – 'I, Samuel Franklin Cody, do swear that I will be faith- ful, and bear true allegiance, to his Majesty the King, and his heirs and successors' – then, when the town clerk's clerk produced his naturalization papers, signed them on the town clerk's back. That done, he snapped to attention and, as the National Anthem rang out, the Union flag was raised and Old Glory lowered, threw a quivering Brigade of Guards salute. Three deafening cheers, led by the town clerk, concluded the ceremony. 'Good old Cody!' yelled the crowd.

He had thought, in fact, of attempting the London to Man- chester himself. The Brock Firework Company, anxious to help, proposed launching polychromatic smoke rockets every few miles to guide him, but he decided to try and reach Manchester from Liverpool instead; £1,000, offered by the owner of an Aintree jam factory, awaited the first aviator to do that. This time, however, perhaps regretting the absence of rainbow-coloured way-points exploding in the sky, he got lost after just twelve miles, flew into a hedge and broke his front elevators. The old Cathedral, he decided, had been smashed and repaired so often it was becoming dangerous to fly. It was time to replace it with something else.

In 2003 TAG Aviation, a multinational operator of business air- ports, signed a ninety-nine-year lease on Farnborough's airfield then embellished it with a new space-age control tower and hangars, and a futuristic wing-shaped terminal. TAG was restricted to twenty- eight thousand movements annually, of which only two thousand five hundred were permitted at weekends; however, since numerous business flights originated in the Middle East – where Sunday is a working day – the company wanted its weekend movements raised to five thousand. Many Farnborough inhabitants, unimpressed, believed they were entitled to a Sabbath relatively free from the scream of jets and the smell of kerosene. (Others, it must be said, adored TAG's twenty-first-century airport architecture, and relished the buzz and glamour of celebrities flying in and out.) A further cause for concern was the suggestion, made on the Farnborough Airport Residents' Association website, that 'there is a serious risk of residents near the airport being killed as a result of a crash'.

The Farnborough College of Technology, situated just two-thirds of a mile from the runway's eastern threshold, boasts a catering school – consistently awarded an 'outstanding' Grade 1 by Ofsted inspectors – with its own training restaurant, the Gallery. When I turned up for lunch a teenage maître d' welcomed me to a light and airy glass-walled room. He was a big, broad-shouldered lad, and seemed amused by the rather panicky way I said, 'Am I the only man today?'

'Looks like you've hit the jackpot, sir.'

There must have been fifty women present, all talking at the tops of their voices. But who might they be?

'A few college staff, but mostly just ladies lunching, actually a lot of them are regulars.' He led me to a table. 'Soup of the day,' he said, handing me a menu and wine list, 'is leek and potato. Enjoy your meal!'

He strode away. I ordered the soup and fish pie from a cheery boy named Sam who, moments later, returned looking solemn. 'Today the First Years are doing the fish. They were *supposed* to make thirteen pies but they only made eleven. Would you like to try the roast, sir? It's pork.'

'Who's doing the pork?'

'The Second Years.'

He was a Second Year himself, and gave the pork a five-star rating. (It turned out to be rather bland and chewy; I gave it two.) After that he brought me the Austrian Dessert Menu offering Vienna bread cream strudel with an almond sauce, apricot gugelhupf soaked in spiced white wine and served with whipped cream, and iced nougat and honey terrine served with meringue mushrooms and a dark chocolate sauce.

I had the nougat. Afterwards, when he brought my coffee (freshly ground) and my bill (£7.50, including a glass of lager), I asked what he planned to do when he qualified. 'I might work on a cruise liner,' he said. 'See the world.' Then I put to him the question that had brought me here: was he worried about the fact that the College lay directly under Farnborough's glide path? He frowned. 'How do you mean?' I said, 'Well, that some day Posh and Becks might come through the roof of your kitchen.'

He said, 'Never thought about it.'

Out in the car park I watched a Cessna Citation, seconds from touchdown, go howling overhead. A middle-aged clerical worker said, 'Occasionally, when something seems to go over really low like that, I get a moment's panic and think, Must move on, must get another job. But I've been thinking that for ten years.' Her friend, a female science teacher, said, 'That goes for me too. I won't stay indefinitely, though; putting a school this size at the end of a busy runway really is tempting fate.'

## Chapter Ten

# Those Magnificent Men

Cody and his Gamages 'non-concussion' helmet.

His shed, out in that isolated corner of Laffan's Plain, stood on earth so damp he could have grown rice in it. There, during the first months of 1910, he worked on a new machine. The Cody Flyer had a wingspan of forty-six feet and stood thirteen feet high; to minimize friction every fitting was plated, every strut French-polished, every plane covered by white camphor-smelling Pegamoid fabric stretched drum-tight. Its two new engines, designed by his friend Gustavus Green to operate a big single propeller, were state of the art – and British. But in June, during an early test flight, he encountered a wind gust so ferocious that, despite his prodigious strength, the control column was torn from his grasp and his shiny new aeroplane tipped up and nosed straight into the ground.

He was rushed home unconscious. Along with his lesser injuries the doctors, grim-faced, diagnosed severe bruising of the brain. A condition like that, if it didn't kill you, could leave you permanently deranged and, for three anguished days, Lela hovered by his bed. Then he suddenly opened his eyes and demanded chicken soup, and shortly after that insisted on being driven, in the pony trap by his groom, Charlie Phillips, to the shed where friends were already rebuilding his aeroplane. This model had modified elevators, and holes drilled in the cylinders to allow a faster dispersal of exhaust gases; when it was ready he declared himself fit enough to fly it.

People called it St Peter's.

It performed well. He took it to various provincial meetings, and was at Bournemouth the day his friend the Hon. C. S. Rolls put on a display. Cody saw the rudder of his Wright biplane snap when he pulled out of a dive then, being first to reach the crash site, wept as Rolls died in his arms. The handsome young aviator who, when not airborne, sold cars built by a young engineer named Henry Royce – mostly to his society friends – was the first Briton to lose his life while flying. (Moore-Brabazon said of him, 'Rolls was the sweetest, kindest, meanest man I ever met. Lunching at the Aero Club he used to bring his own sandwiches and order a glass of water. But I loved him as a brother. We all did.')

The Michelin Trophy, the first prize donated from across the Channel, was intended for the British subject who, in the course of a calendar year, made the longest trip within the British Isles in an all-British aeroplane. The winner would receive a bronze statue and £500 in cash – not much when compared to the kind of prize money on offer from Lord Northcliffe at the *Daily Mail* but, in terms of prestige, enormous. Every pilot in the land entered the 1910 competition, though all knew that, ultimately, it would be between two men. And indeed, as the year ended, only Cody and Tommy Sopwith were left.

Thomas Sopwith

On 30 December Sopwith (a personable young genius who would build the World War One Camel and the World War Two Hurricane and, between the wars, racing yachts that – with him at the helm – twice came within a whisker of winning the America's Cup) took off from Brooklands and, circling the race track for 4 hours 7½ minutes, covered 150 miles. Early next morning, on the last day of the year and in weather so bitter he had to insulate his flying suit with brown paper, Cody began his own attempt. Teeth chattering, quite seriously hypothermic, he spent 4 hours 47 minutes flying round and round Laffan's Plain, and only stopped when, late in the afternoon and after 186 miles, his tank ran dry. Gliding back to earth, beard and clothing coated in ice, he was welcomed by a crowd singing 'Here the Conquering Hero Comes'. The trophy was his.

The following year he won it again – this time beating the British duration record by over sixteen minutes.

The trophies are displayed at the Aldershot Military Museum, situated in the only surviving barrack block at North Camp. (Erected in 1894, it's single-storeyed with a slate roof and tall chimneys, a giant red-brick bungalow possessing the grace and amplitude of a stately home.) I followed a small, nervous-looking man up the ramp and through the door. In a quavery half-whisper he asked the middle-aged civilian woman in charge for the Sexton self-propelled gun. Having given him directions she pointed me towards the Cody Gallery. 'When you get closer you'll hear him talking. Just follow the voice. Cody used to live in the Mytchett Road, you know.' She smiled. 'Just like me.'

The voice (which sounded like a bad parody of Bill Clinton) issued from the head of a life-sized manikin wearing a brown serge suit, black wig, straggly black beard and long black moustaches. 'Feel free to look around, y'all,' it said. As it embarked on a two-minute summary of Cody's life I pondered a pair of Cody's white flying gloves, a white flying cap, a driving licence – 'County of Southampton No. 5125' – and a 'Gamage Pattern "Non-Concussion" flying helmet' belonging, said the exhibit card, to 'S.F. Cody 1910–1913'. I recognized it immediately as the headgear worn, with a very smart suit, in one of his better-known photographs. 'Many of Cody's flights', the card added, 'ended in crashes. And he never strapped himself into his aircraft.' But the display was dominated by the Pegasus bronzes, two magnificent early twentieth-century versions of the winged horse; standing two and a half feet high, they had been presented to Cody by M. André Michelin himself. As I left, the voice said, 'Good to meet you. You go careful now. Byeee.'

'What he planned to do,' said Samuel Franklin John Cody, 'was stick them on his gateposts. One on each. Can you imagine that today? A ton of pure bronze? They'd be nicked within hours.'

The gate, only yards from where I now stood, was familiar to me from photographs of Cody's Union Jack-draped coffin being lifted over it onto the waiting gun carriage. John Cody wore a diamond ear stud and a bead choker, and was quite short, with lively eyes and a round, impish face. His brother Peter was the quieter one, and handsome in a rather grave, old-fashioned way, like a television news-reader from the black-and-white era. Descended not from Cody but from Cody's stepson Vivian, they were exceptionally likeable men; even when I turned up, without warning, at Cody Cars, they were happy to stop what they were doing (looking under the bonnet of a second-hand VW Passat, as it happened) and talk to me. It was they, acting for the family in 1996, who instituted a sale of Cody artefacts and memorabilia at Sotheby's – and the Michelin trophies, each with a reserve price of £45,000, were part of it. No bids of that size were made, so now they're on long-term loan to the Museum. 'Only one other Pegasus bronze was cast,' said John, 'and it was won by Tommy Sopwith. We met him at his house in Hampshire. He told us he'd kept it in his factory at Kingston-on-Thames. During the war they built Hurricanes there so, of course, it was bombed. One of the horse's wings got bent.'

'He was a hundred years old when we met him,' said Peter. 'And quite blind. He ran his fingers over our faces.'

'And he told us,' John continued, 'that though they'd become tremendous rivals, it was our great-grandfather who inspired him to fly. Sopwith crashed his first aeroplane before he was even off the ground so he bought another, and as he was taxiing around Brook-lands to get the feel of it, picked up speed and realized he was airborne. That afternoon he got his brevet licence, that evening he took his first passenger for a spin. Extraordinary man.'

'He told us,' said Peter, 'our great-grandfather had such long hair that, before a flight, he'd gather it up, push a pin through it – just like a woman – and make a topknot. Then he'd pop his hat over that.'

Cody Cars, a modest, single-storeyed wooden structure ('Our motto is "Cody first to fly" ') was located behind 'Vale Croft', Cody's empty, seemingly abandoned house (though still, I learned, owned by the family). I asked about the Sotheby's auction. Some things, it seemed, had done well, the kites going to Drachen in Seattle, for example. 'And we got £14,000 for an early Cody adjustable pro-peller,' Peter told me. 'But there should have been more American involvement, I know there was plenty of interest there. We did lots of interviews, though, John was on Sky so we got some publicity for the family.' An autumnal sun broke through high cirrus cloud that came drifting up from the Channel. It was pleasant standing there, chatting to this agreeable pair. John told me their grandfather Vivian had worked at the Balloon Factory for forty-four years.

'Starting with Army Aeroplane No. 1,' said Peter.

John said, apropos of nothing at all, 'By the way, did you know the front section of the Space Shuttle was developed at Farnborough?'

I gaped at him. Was he telling me that Vivian had also worked *on the Space Shuttle?*

They both laughed. 'Now that really would be something,' said Peter.

'Wouldn't it be just? I said. 'What a way to end the story!'

In March 1911, showing off his Michelin-winning machine at Olympia, Cody was congratulated by George V – who, when they last met, had been Prince of Wales. He said, 'If I were ever to go up in an aeroplane, Mr Cody, I would choose to go with you. Sadly, however, it will never happen. My Ministers forbid me to fly.'

A month later he was in trouble with the Hampshire constabulary. A well-known army entertainer, Captain Arthur Wood, had asked him, as a favour, to drop hundreds of leaflets advertising a forthcoming Aldershot concert. They read:

> A message to you from Cody who flies,
> Re Arthur Wood's venture to open your eyes,
> The date, April 10th, at 8.30 at night,
> These verses are rot, but the show is all right.

They were thrown from the machine by M. Bizion, son of the Empress Eugenie's chef, and, whisked off on a brisk wind, ended up festooning hundreds of gardens. When the householders complained, Cody was warned by police not to do it again and, to ensure no one else seized on what was, actually, a brilliant marketing ploy, Parliament passed a law making it illegal.

The *Daily Mail* proposed a thousand-mile race around Britain, and offered a prize of £10,000. Cody built a smaller, much faster machine with a wingspan of just forty feet.

It became known as St Paul's.

In the summer of 1911, Herbert Asquith, along with his cabinet, three hundred members of parliament (including Winston Churchill) and scores of senior military personnel were invited to a special display at Hendon. As they contemplated the life-sized outline of a battleship sketched on the grass in whitewash, a small aeroplane appeared and started bombing it with sandbags. Claude Grahame-White, a hero of the London to Manchester race and owner of Hendon Aerodrome, was launching a personal crusade called 'Wake Up Britain'. But did the whump of falling sandbags (flour bombs would have been more graphic) force the government to think about its aerial defences? The Military Trials Competition, announced shortly afterwards, indicated that perhaps it had.

Shortly before the Circuit of Britain a trimly bearded figure on horseback cantered across Laffan's Plain and stopped at Cody's shed. It was George V. Being grounded by his commissars may have contributed to a kind of wistful fascination with Cody's new plane, and the idea of what it could do. So, to show him, Cody flew a few low-level circuits. The King wished him luck for the coming race, shook hands and rode away.

It would cover one thousand and ten miles exactly, with stops in various specified places. Of the twenty-one pilots who showed up for the start at Brooklands on Saturday 22 July 1911, nine were British, the rest well-known aces from France, Austria and Holland. Though temperatures stood in the nineties a crowd of thirty thousand (paying the local citizenry a penny a glass for water) saw them off. First to go was a Frenchman, who immediately crashed. Next were two machines entered by Bob Blackburn of Leeds. The first crashed – the pilot, Conway Jenkins, was unhurt – while the second, possibly flown by a figure known as Peter the Painter, made a forced landing at Luton. Of three machines sent by Sir George White, a millionaire tram manufacturer from Bristol, the first crashed while taking off, the second never left the ground, while the pilot of the third had had his licence revoked for making a recklessly low pass across the Henley Regatta (scattering the boats and sending spectators diving for cover). The official halt that night was at Hendon, where Cody arrived without incident.

Six more crashed on Sunday on their way to Harrogate where, before a hundred-and-fifty-thousand-strong crowd, Jimmy Valentine, the first Briton to land, received a silver tea service from the mayor. Cody, having diverted to Rotherham with radiator trouble, touched in Harrogate with a leaking fuel tank. At 4.51 a.m. on Monday, while the others slept, he took off for Newcastle, lost his bearings in heavy fog, finally flew into clear air over the village of Brandon. There, landing to ask directions, he collided with a wall. Word of his arrival spread like wildfire, excited folk came from miles around, children were given a half-day off school. Cody was invited to stay with Henry Cochrane, owner of Brandon Hill colliery, and, awaiting new parts, saw otters in the River Browney, had his spirits lifted by the quiet beauty of the Durham Dales and, of course, talked. A photograph of the time shows a community in holiday mood. There, in the background, stands the great flying machine, in the foreground small girls in straw hats and summer dresses are seated on the grass (perhaps making daisy chains for the famous aeronaut) while their fathers and brothers, wearing cloth caps and Sunday suits, crowd around the aeroplane; Cody may be hidden from sight but they're plainly hanging on his every word. (In Durham, in 2005, I met an elderly woman who told me her grandmother, on that red-letter day, had baked a cake for Cody.)

Later in the week, as he completed his repairs, news arrived that the race had been won by a Lieutenant de Conneau, who worked

Cody, during the Circuit of Britain Race, was forced down on
Brandon Hill. Those who came from miles around to see the legendary
aeronaut, and his great machine, included children from the
local school. They were given a half-day off.

for Blériot and had, that very day, received a cheque for £10,000
from Lord Northcliffe.

It was generally assumed that Cody would head for home. He
decided, instead, to complete the course on his own.

To do it properly, though, each specified stop would need to
be visited. He'd missed Newcastle so, on Thursday, he headed
there. From Edinburgh, where he lunched, he flew to Stirling where,
before a huge welcoming crowd, he ran into a fence. At Glasgow he
awarded himself a weekend off. On Monday he called at Carlisle,
stopped at Whitehaven for breakfast and a shave (the barber's shop,
Cinammon's, placed his tuppenny fee in a glass reliquary box for
everyone to see), moved on to Manchester and Worcester, reached
Bristol on Wednesday, spent Thursday night in Weston-super-Mare
(*The Klondyke Nugget* had once played there, and his fans, to welcome
him back, draped his aeroplane in bunting) then flew on to Exeter.

On Saturday, facing a 35 mph headwind, he got to Shoreham via a touchdown on Salisbury Plain, late that afternoon, doing 90 mph with the wind now behind him, he finally crossed the line at Brooklands – the only all-British machine to have done so.

He received a rapturous reception. It had been a performance so dogged and brave the country hailed him as a hero all over again. Even the press now treated him with deference. The *Aeroplane* wrote, 'He himself is a self-made man without friends or money to back him, yet he is as true a gentleman as ever lived – honest, straightforward, kind-hearted, a thorough sportsman, a magnificent flyer, full of pluck and as straight as they make them.'

The *Morning Post* declared, 'He is a big man but deserves to be called, in the best of senses, child-like.' There was a reference to Cody's love of bread-and-butter pudding, but that wasn't what they meant.

The 1965 movie *Those Magnificent Men in Their Flying Machines* was inspired by the 1911 Circuit of Britain race.

Lieutenant de Conneau was honoured with a ten-course banquet at the Savoy. Cody had tea and scones – some historians say it was actually potted-shrimp sandwiches – at the Blue Bird Restaurant, located in a Brooklands hangar, and paid for out of his own pocket.

Brooklands, the world's first purpose-built motor-racing circuit, had been opened in 1907 outside Weybridge, a mere hop and skip from Farnborough. Cody dropped in often, liked the place, and the jaunty, innovative, technically brilliant, speed-mad characters one bumped into there, individualists like Louis Zborowski, a Polish count who would, one day, build a giant, awesomely fast car named Chitty Bang Bang (with a chain-drive Mercedes chassis, a duck's-back body, a six-cylinder Maybach aero engine and, in place of an exhaust, a chimney from a wood-burning stove). While the crowds flocked to see stars like him (and Cody), those in charge at Brooklands – the Automobile Racing Club Committee – knew, when it came to speed and spectacle, that aviation was the coming thing. Very soon they would plant an aerodrome plumb in the middle of their severely banked, 3¼-mile concrete track but, first, they announced a prize of £2,500 for the first person to *fly* around it before the end of 1907. Their challenge was accepted by Alliott Verdon Roe, a thirty-year-old zealot who was to become Cody's chief rival for the first-flight crown.

Born in Patricroft, Greater Manchester, son of a wealthy GP, he had gone to sea as fifth engineer on a British and South African Mail Company ship and, down in the Southern Ocean, watched albatrosses. 'If a bird could glide like this,' he mused, 'why should not a man do likewise?' He came ashore, turned the stables of his brother, Dr S. Verdon Roe, a Putney GP, into a workshop, created a model plane that won a competition at Alexandra Palace (soaring so far it had to be restrained by safety nets) then started on a full-sized aircraft. Accustomed to dealing with giant coal-fired engines, he now had to think in gossamer weights, and construct parts – king posts, flywheels and U-bolts – light enough for a baby to lift. It was coming along nicely when the Brooklands Committee announced its prize.

He tore out to Weybridge on his bike. A wooden notice – it's there still – said 'Clerk of the Course To Whom Apply In All Matters'. For reasons not fully understood the clerk, a Mr E. de Rodakowski, took a profound and immediate dislike to Alliott. He, unaware of this, asked permission to bring his unfinished aeroplane to Brooklands. De Rodakowski had, in fact, designed the racetrack but, since it was his directors' declared policy to encourage aviation, he had little choice. Alliott got his permission through gritted teeth.

He trucked the machine over, erected a small shed and painted AVROPLANE above its door. De Rodakowski took to loitering balefully outside, uttering imprecations and abuse. Alliott, subsisting on kippers and dates, laboured on.

One morning in May, 1908 de Rodakowski's simmering rage went critical. Saying this was supposed to be a racetrack, not a repository for lost causes, he insisted his shed be moved to the far end of the paddock and painted green to make it indistinguishable from the trees. Also, the machine could only be brought out at dawn and dusk, when nobody would be around to see it. Alliott, just days away from his first trial, agreed.

On 8 June 1908, the tiny Avroplane was ready and, as if absorbing some essential essence of albatross, flew without even being asked. Years afterwards Sir Alliott Verdon Roe recalled what happened. 'I had been taxying along the track at Brooklands when I realized that I was clear off the ground. I was flying for the first time. My flight in the air was over a distance of a hundred and fifty feet, and I made a perfectly smooth landing. As I had not arranged for official observers to be present it was not officially recorded. Luckily, however, both the head carpenter and gatekeeper had seen me, and I was subsequently to obtain signed statements from them.'

It happened a clear four months before Cody got airborne, so was Alliott, as he always claimed, the first man to fly in Britain? The Aero Club loftily decreed it had been a tentative hop, not a real flight at all – a finding with which Geoffrey de Havilland angrily disagreed – and, ignoring Cody, handed their Licence No. 1 to John Brabazon for a five-hundred-yard sortie he'd made, in a French-built Voisin, at Leysdown on 2 May 1909; Alliott went on to achieve *his* first officially approved flight in July of the same year. (Yet the story of Tommy Sopwith rising into the air without realizing he'd even left the ground inclines me to believe that Alliott, on 8 June 1908, made a spontaneous ascent too. It's actually a pleasing idea – the Balloon Factory's Cody and Dunne, backed by the might of the British government, beaten by a single, solitary, thoroughly obsessed young man.)

When word reached him that the Wright Brothers were staying at Pau, near Biarritz, he cycled to Southampton, took a ferry to Saint-Malo, rode the four hundred miles to Pau, peppered the astonished Americans with questions, then did it all in reverse, arriving home again to resume his search for cheap premises with flat land adjoining. Having no wish to return to Brooklands he had initially sought

Alliott Verdon Roe. He built his first aircraft at Brooklands and, some believe, became airborne four months before Cody.

permission to join Cody on Laffan's Plain but the War Office, shaken by the prospect of *two* dysfunctional tenants on its territory, refused. So did the authorities controlling Hackney Marshes, Wimbledon Common, and the spaces adjoining Wormwood Scrubs Prison. Finally, for a peppercorn rent, the Great Eastern Company let him have a pair of disused railway arches on Lea Marshes in Essex.

The volunteers turning up there to help included a hunchbacked Cambridge engineering graduate named Howard Flanders, the young Egyptologist E. V. B. Fisher (who, according to Alliott, had 'the face of a poet') and a Miss Mildred Kirk from Derbyshire. They lunched off bread and marmalade – money being short – and, in winter, when the marsh froze, went skating. That, however, all changed when Alliott's elder brother, Captain Humphrey Verdon

Roe, appeared. A veteran of the Siege of Ladysmith, he had inherited, from an aunt, the failing Bulls-Eye Trouser Braces Company – facing closure until he came along. Now, thanks to his boundless energy and bottomless self-belief, Bulls-Eyes held up the pants of half the men in England; and Alliott soon learned that Humphrey had plans for him too.

At the time he was building a second machine (named the Bulls-Eye) from pine and muslin-backed packing paper, its undercarriage a set of backward-castored bicycle forks, the skeletal fuselage so flimsy all one saw in flight were twin sets of triple-decked wings. These resembled a pair of giant chasing butterflies with Alliott – wearing his cap back-to-front – somehow suspended between the two; from time to time, sounding like a car crash duplicated instantly by its echo, they chased each other right into the ground.

It made for a great spectacle, and crowds of East Enders – who christened him 'The Hopper' – came to watch. Leyton Council, however, put him on a public-nuisance charge. First, though, he had to be caught in the act, so he flew at the crack of dawn, lifting off as the first grey fingers of light stroked the marsh. The bailiffs only got him after he had crashed, serving their summons as he sat, stunned but elated, amidst the wreckage of an aeroplane that had just climbed to 900 feet. It had been the best flight of his life.

Soon afterwards, at the Blackpool Air Show, he won £75 in prize money then crashed again; a day later, in Blackpool's Holy Trinity Church, he married Miss Mildred Kirk.

Despite his ferocious assiduity – he always worked in a dark worsted suit and high starched collar – he was a gentle, modest soul, remaining largely indifferent even to his knighthood and fortune. 'You couldn't help liking him,' recalled George Halse, one of his mechanics, while Halse's brother Arthur returned from a long period abroad to find Alliott 'famous but unchanged'. Yet, towards the end of his life, distressing and embarrassing his friends, he became an outspoken supporter of Sir Oswald Mosley.

\*

At Brooklands, in the 1907 Club House with its pictures and mementoes of dead drivers, its roped-off offices, lounges and billiard rooms, the demeanour of visitors was awed and whispery; British motor-racing fans wandered around like Marxists in the Kremlin (gazing, wide-eyed, upon Stalin's bath, Lenin's bed). I asked one of the elderly male volunteers who run the place where I might find Steve Devereux. He said, 'In the Balloon Hangar, probably. Go left out the door and past the Petrol Pagoda.'

It had rained all week, but on this late summer's day in Surrey's hinterlands the sun was hot, blackbirds sang, a lark joyously ascended. Brooklands, I saw, seemed to occupy a deep grassy concavity. A wall, looking like the excavated foundations of an old city, went sloping down its side to paddocks in which marsh marigolds, sweet violets and wild daffodils bloomed. There was nobody about. Then I turned a corner and found, parked massively ahead, a Concorde in full British Airways livery. A blonde woman, who thought she was alone, placed her hand on its nose-wheel stanchion and stood quietly, as if recalling a particular journey, or someone she'd made it with; when she'd gone I did the same, my own memory being of a trip to Bahrain on one of Concorde's first-ever commercial services and, as the bulkhead machmeters indicated we were through the sound barrier, a gang of Britain's top businessmen whooping and yelling like schoolboys.

Around the next corner I came upon a field in which more retired airliners stood. They included a Vickers Viscount and a VC10, both Brooklands-built, the latter a cloud-hopping royal yacht made by British craftsmen for Sultan Qaboos bin Said, ruler of Oman. It had a forward stateroom with an exquisitely curved laminous wall and a pair of great throne-like facing armchairs. Further aft spacious wood-panelled sleeping chambers contained beautiful wooden bedroom suites. A group of small children were leaving as I arrived. One, wide-eyed, said to me, 'They had *showers.*' Outside again I took another look at that ancient city wall, and realized I was surveying the remains of the Brooklands' racetrack with its terrifying one-in-four gradient. The surface, six inches of Portland cement laid a hundred years ago, was now so rough and discoloured it had taken on the appearance of granite.

I came to an echoing hangar filled with vintage planes. Another ageing volunteer told me the Balloon Hangar was next along. This, he said, was the Wellington Hangar, named for a bomber that had crashed into Loch Ness in 1940, been located by American monster hunters in 1976 then raised, brought here and put back together again; an almost complete Wellington stood before me today. 'Are you a retired flying man?' I asked him. He said, 'Actually, I'm a retired train driver,' and showed me a lovely little wooden amphibian built by a team of Brooklands women in 1919. 'The wings are spruce, the hull is elm,' he said. 'One landed on the Thames at Westminster at low tide, put down its wheels and taxied onto the mud right under the House of Commons!'

I had come to meet Steve Devereaux at the urging of a journalist friend. 'He'll give your book a bit of balance,' she said. 'Brooklands in many respects was just as important as Farnborough, and Steve's the chap to tell you about it.'

He turned out to be tall and white-haired, with blue eyes, pink cheeks and a weathered face, one of those active, interested men who remain pretty much unchanged by age. His workshop stood in a shadowy corner of the Balloon Hangar – a modest-sized shed once containing a supersonic wind tunnel. 'Used in early development work on Concorde,' said Steve. 'Don't forget the idea for that sprang from a meeting at Brooklands.'

Today the hangar displayed a small blue glider that had belonged to the legendary Thirties racing driver – and Brooklands darling – Prince Birabongse Bhanuban of Siam. (A photograph showed a slight, dapper, smiling man with perfect teeth and slicked-back pomaded hair.) Steve pointed out a tiny window in the rear of the cockpit roof. 'He always flew with a Pekinese sitting on his shoulder, and had that put in so the dog could see the sky. During the war he became a gliding instructor for the Air Training Corps, but this machine was kept for his private use at home in Cornwall. Several years ago his widow paid us a visit, and was amazed to find it here. She told us how Prince Bira used to phone from *miles* away asking her to come and fetch him *yet again*. She'd chased that thing all over the West Country.'

When I told him of my interest in early Farnborough he jumped onto a chair and took from a shelf a photo of a giant hangar with, parked outside, a six-engined triplane so immense its pilots, seated high in the nose, probably got vertigo just from looking down at the tarmac. 'Taken in 1917,' he said. 'The hangar you'll recognize.'

It was the Balloon Factory.

'This is what they got up to after Cody died. That aircraft, the largest ever built at Farnborough, was the plane that would win the war. It was designed by a man named Tarrant Tabor, and was meant to carry so many bombs the Kaiser would soon be begging for mercy. But, when it was wheeled out of the Balloon Factory, they needed a whole day just to get the engines started. Then, instead of rising into the air, it tipped over on its nose. The press ridiculed it so the War Office cancelled it; and poor old Tabor's super bomber never even got off the ground.'

I wanted coffee, and asked if, by any chance, the Blue Bird might still be around. He smiled, and said the hangar in which it stood, on the far side of Brooklands, had been demolished long ago. 'There's a Marks & Spencer's there now, it's all shops.'

So we drank our coffee in the airy Club House restaurant. He told me he had first come to Brooklands as a seven-year-old in 1930, brought by a rich uncle who always rode a motorbike down from Wiltshire. 'He loved the cars, but what interested me, right from the start, were the aero engines. And that never changed. I spent the war working on them in the RAF, in 1950 I joined the Vickers tool room at Brooklands. Flying ended here in 2000, when the strip was damaged by floods and frost. But I'm still around.'

Had he ever met Alliott Verdon Roe?

'No.' And that was all he wished to say about Alliott. He was, though, happy to discuss an Avroplane replica to be built at Brook-lands (yet without using bamboo; a key component of the original machine and, of course, indispensable to the pioneers – Cody used up more bamboo than a tribe of pandas – it's now held to be too dangerous to work with). I said Farnborough was planning a replica of the first Cody machine.

'I've heard about that,' said Steve. 'But it'll be static.'

'Meaning it won't fly?'

'Exactly.'

Then I learned about the rivalry between the two places. 'From 1907 to say 1977,' he said, '*unlike Farnborough*, we were producing aircraft non-stop: a hundred and eighty prototypes, and twenty-eight thousand production models – including every Viscount, Vanguard and VC10 ever built – first got airborne at Brooklands.'

Sir Barnes Wallis, he continued, spent his career here, devel-oping the Wellington bomber, along with the bouncing bombs dropped by the Dam Busters, also Tallboy (which sank the *Tirpitz*) and Grand Slam, the largest conventional bombs used by the Allies in World War Two; the US bunker-busters were direct descendants.

The other thing that Brooklands did was teach enthusiasts to fly. 'Of the hundred licences first issued in the UK forty went to people who had learned here.'

As for Farnborough – well, what could he say? 'When Frank Whittle came up with the idea of jet propulsion he was summoned there for an interview. He saw a man named Griffiths who decided that Whittle was mad, and his newfangled idea was too silly for words; thanks to him, thanks to *Farnborough*, the jet engine was put back by two years.'

Yet it seemed to me that Brooklands, after decades of furious activity, had lapsed into an almost narcoleptic state while Farn-borough – with its glitzy new airport, its hundred-and-eighty-acre business park (set on the old Factory Site) and Cody Technology Park (home to QinetiQ, known for its defence-research work and modish misuse of typography) – was firing on all cylinders. I said goodbye to

Steve and set off for Weybridge station, on the way recalling once meeting an old lady in Auckland who claimed that, as a child, she had been befriended by 'the world's first woman pilot'. Brooklands had somehow been part of that story, so next day I switched on my computer and activated certain search engines. And they came up with something so curious it's worth repeating here.

Hilda Hewlett, a London vicar's daughter and trained nurse, decided to take up flying after watching A. V. Roe at the 1909 Blackpool Air Show. She enrolled at Brooklands and, graduating with 'Ticket No. 122', became Britain's first female pilot. Aged forty-six, attractive with dark hair and a prominent nose, she continued her training in France then, in the summer of 1910 – now calling herself Grace Bird and accompanied by the Gallic ace Gustave Blondeau – returned to Brooklands. There she gave lessons, one student being her son, Francis, a naval officer – who became the only military pilot on earth taught to fly by his own mother – another an eager, fresh-faced Tommy Sopwith, then aged just twenty-one and riding an aeroplane for the first time in his life. Two circuits of the track cost him £5, while back on the ground and near-delirious with excitement, he placed an order for his first machine, a little 40 hp Avis that set him back a further £650.

Grace Bird's husband was the well-known romantic novelist Maurice Henry Hewlett. (A painting in the National Portrait Gallery shows him to be greying, bookish and bored-looking.) His 1898 saga *Forest Lovers*, about chivalry in the Middle Ages, was a bestseller extravagantly admired by Marie Stopes. She had just begun her classic book *Married Love*, which postulated that the miserable state of many British marriages was due to a profound ignorance of sex; indeed, in 1914 – the year Grace 'politely' left Hewlett – Marie Stopes' own marriage ended unconsummated. (Later, in what seems to have been a happy union, Stopes wed Captain Humphrey Verdon Roe.)

In 1913 Grace and Blondeau, realizing the Great War was approaching, set about manufacturing aircraft. Their Bedfordshire plant employed seven hundred people and produced, under licence for the Royal Flying Corps, ten different types of machine. A decade later, aged sixty-two and weary of 'crowds, convention and civilisation', she boarded a steamer for New Zealand and never returned.

## Chapter Eleven

# Colonel Cody,
# Pillar of the Establishment

Cody after winning the Michelin Trophy.

Throughout the kingdom various minor race meetings were held. In one Cody beat his old friend F. P. Raynham (allegedly the first pilot ever to recover from a spin – and find, for himself, a significant little niche in the history of aviation). But Cody longed for a big Blue Riband event, some great project that would extend him to his limits, and, late in 1911, the War Office obliged. It announced a competition for a new military machine which, he gleefully told his friends, would be required to climb to 1,000 feet 'in less than four minutes, carry half a ton, land in the most diabolical ground' and be so logically constructed that repairs could be carried out by 'a carpenter or house-builder'.

The ruling that British aviators must use British engines had been relaxed and, having ordered a big water-cooled, copper-jacketed 120hp Austro-Daimler, Cody began work on the monoplane it would drive. It was only half finished, though, when the engine arrived. Anxious to see it in action he installed it in the 1910 biplane in which he'd won his first Michelin, started it up and watched it blow the biplane's tail clean off. (Here I see Cody dancing a gleeful little jig.) He replaced the tail then, calculating that he could increase the payload of his old machine by a factor of five, fitted four new tractor seats, in pairs (the rear pair, set higher, he called the Dress Circle).

One afternoon he took his youngest stepson, his niece and two of his loyal – though probably unpaid – helpers, Mr Tudgey and seventy-year-old Mr Wackett (seated in the Orchestral Stalls) for a high-speed jaunt around Hampshire, and was so pleased with the way it performed he began using it for public joy rides. The cost of a flight with Cody was £2 per person per time.

Geoffrey de Havilland regarded Cody as 'the most colourful' of the people he knew at Farnborough. Cody, he wrote, was 'an adventurer', 'an individualist', a 'loveable' giant who 'must have weighed all of fifteen stone . . . his naturally dramatic appearance emphasized by his small goatee beard, waxed moustache and long hair which fell over his shoulders when . . . not tucked in under [a] cowboy hat.' De Havilland first met Cody when he joined the Balloon Factory in 1911 and, years later, was still angered by his banishment to Laffan's Plain. 'I often used to fly over there to talk to this fascinating and brilliant character. His enthusiasm and practical, down-to-earth approach were infectious.' Cody might have had not a jot of engineering training, yet he was able to judge 'the strength of things and the fitness of detail parts purely by eye. He used bamboo extensively . . . and fixed the steel bracing wires by tying them in knots with his enormously strong hands.' Cody's methods might be 'makeshift', but they produced innovative, outstandingly successful machines.

The sound of de Havilland dropping in (noted an anonymous diarist who happened, once or twice, to be present) was 'a faint and pleasant purr' in the sky. He was a quiet, thoughtful, donnish man, perhaps a little aloof yet, despite their rivalry and difference in ages – Cody, at forty-seven, was sixteen years his senior – there was mutual regard (de Havilland, for a start, probably couldn't get over the terrifying array of ditches and declivities from which Cody routinely operated). Cody, having maybe showed the young aeronaut some item of interest in his workshop, made strong tea. Then the two exchanged news and opinions (neither was interested in gossip). When, finally, his visitor hopped back in his plane and purred off again Cody would stand – hat at a rakish angle, rocking gently on his heels – with a hand raised in farewell.

By the spring of 1912 most of the world's major manufacturers were preparing for the British Military Trials; Whitehall, suddenly realizing that the next war might, after all, be fought with aeroplanes, offered a £5,000 prize plus a government contract to the firm entering the most reliable and – this had to be said sotto voce – *cheap* heavier-than-air machine. (George V, inspecting the dozen ragtag aircraft of the new Royal Flying Corps, tactfully didn't ask how the French Armée de l'Air had managed to acquire three hundred.) Then, out of the blue, Cody was challenged to a race by the Balloon Factory: one of their spanking new de Havillands versus one of his old biplanes (but fitted with the Austro-Daimler). Both aviators were equally puzzled as, lined up side-by-side, they were ordered to start their engines then head off down a precise 1½ mile course. Cody, doing 70 mph, beat de Havilland by a clear twelve seconds.

They were never told the point of the competition, but Cody

now determined to enter two machines for the Military Trials: his new, unfinished monoplane, and the biplane in which he'd raced de Havilland, and which – he was, as usual, chronically short of money – he began using for lessons. But a student from the navy misjudged a landing and hit the ground so hard he and Cody were thrown out. Both suffered severe bruising, while the aircraft, spinning into trees beside the Basingstoke Canal, was reduced to matchwood. Only one biplane, the Circuit of Britain machine now powered by the Austro-Daimler, remained serviceable but, shortly afterwards, a student from the army crashed that too. Next he fitted the Austro-Daimler, miraculously undamaged, to his sleek new Trials monoplane which, with its interestingly curved body – from certain angles it had a faintly surreal look – and partially enclosed cockpit (his legs stayed warm), impressed all who saw it. *Flight Magazine* reported that the new Cody cruised at 80 mph, and that virtually any part of it 'could be repaired by a blacksmith, which might be . . . most useful in war'; Cody, delighted, reckoned he could actually win the competition.

Of the thirty-one entered for the Trials only five were all-British, thirteen more were British-built with foreign engines – Gnomes, Chenus, Anzanis and, of course, the Austro-Daimler – and the rest were manufactured in France and Germany. The Trials were due to start on 2 August, and on 4 July Cody was seen to be doing well, moving 'terrifically fast'.

At about that time the anonymous diarist was given a spin, Cody flying terrifically fast through Scotch mist at a height of about five feet, evidently following Laffan's Plain's eccentric contours by instinct. Suddenly he said, 'Now I'll show you!' and, the diarist wrote, 'pulled back the cloche, and at the same time opened out full. The result was extraordinary.' Several days later, during a high-level run, engine failure forced him to glide back to Laffan's. Just seconds from touchdown, as he passed soundlessly over a herd of cows, something – perhaps his rushing shadow – stampeded them. One blundered directly into his path and was clubbed by the undercarriage. It died

Cody in the monoplane he built for the Military Trials – then wrecked
when colliding with a cow on Laffan's Plain. The partially enclosed
cockpit kept his legs warm.

instantly,* while Cody, bruised and shaken, staggered uninjured (his
luck holding still) from the wreckage knowing the Trials would begin
in three weeks and he had no aeroplane.

So, starting from scratch and working around the clock, he built
another, a four-seater biplane with reinforced wings, twin rudders
and a chain-geared, twelve-foot Chauvière propeller powered by his
battered but indestructible Austro-Daimler. After a hasty test flight
he proceeded – at 73 mph – to Larkhill aerodrome, near Salisbury.
There he inspected the opposition, saw that, with one exception,
all were machines entered by companies – for example, the British
& Colonial Aeroplane Co. Ltd (owned by Sir George White, the
Bristol tram millionaire) and the Société Anonyme des Aéroplanes

---

* When its owner sought damages through the courts, Cody's lawyer claimed
that the cow, chronically depressed and in suicidal mood, had thrown itself in
front of the machine. The judge awarded the owner £18.

Borel – with big budgets and plenty of trained personnel. The only private individuals bold enough to enter were Cody, and a German named C. E. Kny.

The Trials lasted a month. Every aspect of performance, handling and construction was rigorously checked: competitors had to stay aloft for three hours, attain a height of 4,500 feet, climb and descend at specified rates, demonstrate their highest and lowest speeds, fly in a strong wind, land in a ploughed field, take off from long grass or clover, glide far enough for a gliding-angle gradient to be established, taxi in figures of eight to test ground steering, allow the engine to 'tick over' while stationary, dismantle the aircraft, put it in a truck, drive for miles then reassemble it. Marks were awarded for cockpit comfort ('protection from the wind'), simplicity of controls, field of view, undercarriage design, overall weight and the shortest landing. (Cody, cunningly, had fixed a chain to his skid which he yanked up before take-off, and dropped the instant he touched down; working like a hoe, it brought him to a juddering halt in just thirty yards – a distance no one else could match.)

The corporate crowd stayed at Salisbury's best hotels; Cody's team – who had cycled the sixty miles from Farnborough (old Wackett rode a tricycle) – slept in the hangar. So did Cody.

Just eleven aircraft finished. Several crashed – the Avro turning a complete somersault – but with just one fatality: R. C. Fenwick, flying a Mersey monoplane. (It was thought that the rubber soles of his shoes, made slippery by wet grass, slid off rudder bars that lacked stirrups or end-stops.) The other non-finishers were simply retired.

On 30 August 1912 the War Office published the results. There were two categories, Foreign and Home, and they read as follows:

(A) Prizes open to the world for aeroplanes made in any country:
    1. (£4,000) To S. F. Cody for Cody biplane, British, No. 31.

(B) Prizes open to British subjects for aeroplanes manufactured
    wholly in the United Kingdom except the engines:
    1. To S. F. Cody for Cody biplane, No. 31.

The King addressed a congratulatory telegram to 'Colonel Cody' and, from then on Cody (infuriating the holder of every genuine colonelcy in the land) adopted the rank and title – and encouraged everyone else to do so too.

In London the new Colonel Cody met the original Colonel Cody. Some feared that Buffalo Bill, knowing about the small deceptions practised by Sam during his stage career, might hurl charges of fraud and criminal impersonation at him, or that there would be a titanic clash of very large, highly combustible egos. In fact, the two Americans got on so well they made plans to meet again.

The Royal Aero Club finally awarded him its gold medal. He had become a member three years earlier, but, even after his royal promotion, nobody there ever called him Colonel.

Staff and fellow-members alike had their own name for Cody. They called him Daddy.

Now he sought the kind of wealth that would match his fame and reputation. A new enterprise, Cody & Sons Aerial Navigation Ltd, was announced with some excitement, and a brochure issued. It contained a picture of Lela seated, in long skirts, on a trapeze below an airborne Cody kite, and offered customers a choice of three biplanes ranging from £750 to £1800. He was plainly inspired by Sir George White (recalled by Penrose as 'a dapper little Edwardian baronet, fierce in manner') who, in his splendid Filton factory, now produced a hundred machines a year. But Cody wasn't cut out to be a tycoon, in fact seemed happiest working away in his shed with Tudgey and Wackett; Cody & Sons never got off the ground. The War Office bought his winning Trials machine (and crashed it) then, having promised a large order as part of the Trials prize, asked only for a single replica – and crashed that too. (Entirely rebuilt, it's now displayed in London's Science Museum.)*

* Where it occupies a position of honour, hanging from the ceiling beside Amy Johnson's Gypsy Moth, and the Vickers Vimy aboard which, in 1919,

Cody grew bitter. He told an interviewer, 'I built the first aeroplane that flew in England, and I built the first man-lifting kite which led the way for the aeroplane abroad, and I put wings on the kite before the Boer War started. My machine was built under the auspices of the War Office and entirely my design. They were of the opinion that it was no use.' Then, displaying the self-pity that was among his least attractive traits, he complained about 'the neglect, the scorn, the contumely which has been heaped upon me. I hadn't even a shed with a floor in it, so I had to work on muddy ground, and it's a swampy place where my *present* shed is.' (When it rained, as he well knew, the whole of Laffan's became a swamp, yet it was entirely his decision to stay.) Steeped in bathos he carried on, 'I continued for three years and allowed the world to go on against me,' then, in a typically abrupt mood change, crowed, 'But I still came up and beat them!'

The War Office, though, couldn't have its Military Trials winner carrying on like this; it might affect sales abroad. (British machines were now much in demand.) It's my belief that, quietly, they readmitted him to the Balloon Factory and, without urging him to quit his shed (which, perversely, he loved anyway), granted him a toehold, a cubbyhole, a postbox, *a presence.*

He continued to enter competitions and win trophies – receiving one from the dainty hands of the actress Marie Tempest, then appearing before packed houses in Somerset Maugham's *Mrs. Dot.* (Famed for a rather arch combination of charm and roguishness – which she doubtless used on Cody – Tempest would, in 1925, inspire Noël Coward to write the part of Mrs Bliss in *Hay Fever* for her.)

***

John Alcock and Arthur Brown made the first non-stop flight across the Atlantic – thus winning a £10,000 *Daily Mail* prize on which Cody had set his heart. His biplane is fitted with tractor seats and twin kite-like tails, while twenty-six quarter-inch-wide wooden ribs are laid, with extraordinary delicacy and exactitude, across the upper and lower surfaces of each wing. Beside the big, lumpy Vimy with its two giant Rolls-Royce Eagle engines, Cody's Military Trials machine looks positively spectral.

Cody's elegant Water Plane, on Eelmore Flash.

In 1913 Lord Northcliffe proposed holding a 'water plane' race around the coast of Britain. Cody, noting the £5,000 prize, posted off his entry and, using his Military Trials money, built a giant 'hydro-biplane' with sixty-foot wings, outer pontoons and a central float. (He filled these sleek racing shells – made by a Mr Harmsworth at the Ashvale Boathouse – with family and friends, roped them to a fast motor launch then took them on a thrilling dash along the Basingstoke Canal.) The finished plane, weighing over a ton, created a wash that scattered the ducks and tugged at the reeds yet, at rest, resembled a giant dragonfly barely breaking the water's skin. The most elegant, aesthetically pleasing machine he'd ever made was also, ironically, the one that would kill him.

He kept it moored on a pond called Eelmore Flash – a widening of the canal close to his shed – and used to pull it ashore at Laffan's Copse. The land there was all part of Laffan's Plain: today Laffan's Road passes the Army Golf Course, and a Laffan Track marked (and continues to mark) the boundary of the airfield.

One day, hopping into a taxi at Farnborough station, I told the driver I wanted to see Eelmore Flash. Did he know where it was? 'Course I know,' he said. He seemed to know a bit about Cody, too, and referred to a vanished pub called Cody's Tree. 'Used to drink there; I'm a bit of a historian myself, to be honest,' he said. Well, that's good, I thought.

On Farnborough Road he pointed out the stately old Balloon School, built by Royal Engineers in 1907. Painted white, set behind a modern high-security fence, it had several vintage jets parked in its front yard, and the nose of a Canberra bomber (cockpit included) mounted and stuck, like a stag's head, to an exterior wall. He nodded at it. 'Today that's the Air Sciences Museum.'

'Let's pop in,' I said.

'It only opens at weekends. Lord Trenchard is supposed to have created the Royal Air Force there in 1918.' We passed Government House Road containing the residence (occupied by the Aldershot Garrison's Commanding Officer) from which a young George V had ridden out to meet Cody, then turned right along tree-lined Fleet Road – a sign said 'Troops Crossing' – which offered occasional brief glimpses of a narrow, leaden-looking canal. A squad of soldiers doubled by. I said, 'Look at them, they should still be at school.' The driver shrugged. 'You'd probably find one or two Iraq veterans among that lot.' Soon afterwards I saw futuristic new buildings, and signs saying 'Cody Gate' then 'Cody Technology Park'.

'What on earth are those?' I asked.

'It's the new airport entrance. These days they're naming everything in sight after Cody. Like they all just woke up and remembered who he was.'

'Better late than never,' I said. 'So where's Eelmore Flash?'

'We passed it.'

I stared at the back of his shaven head. 'What do you mean, *passed* it?'

'Five minutes ago.'

'But that's where I wanted to go.'

'No, mate, that's what you wanted to see. It's what you said back at the station: "I want to see Eelmore Flash" – which you did. I spotted you looking in my mirror.'

This was extraordinary. 'What I meant, obviously, was that I wanted you to *take* me there. And, when we arrived, to stop and let me out.'

'You didn't say nothing about *stopping*,' he grumbled then, sighing heavily, doubled back and halted by an inlet so spacious you could have anchored an Empire flying boat in it. The view, though, was swamped by the proximity of the runway lights, brilliant blues, golds and greens which went twinkling off in a tremendously celebratory and glamorous fashion. A short distance away, at the water's edge, a youngish man with protruding ears and thinning, sandy hair sat sketching on a folding stool. We got talking. Cody's connection with this particular spot seemed not to interest him as much as its flora and fauna. 'It's got the biggest range of aquatic plants in the country,' he said. 'They even made it a Site of Special Scientific Interest.'

I looked at him in surprise. 'You mean the Flash?'

'No, no, the canal. You know the story?'

He told it succinctly. The canal went from Basingstoke in Hampshire to West Byfleet in Surrey via twenty-nine locks and a twelve-hundred-yard tunnel that passed beneath the hill at Butterwood. After the war it began drying up, finally dwindled to little more than twenty-nine duck ponds joined by a muddy trickle (its owner, a businessman, was reduced to selling fishing tickets). Then, one morning in the Sixties, torrential rain caused it to burst its banks and flood Farnborough's runway – *on the opening day of the Air Show*.

Yet, despite the air marshals and ministers splashing around, puce-faced, with their trousers rolled up, the state chose not to intervene. Rather, in the British way, it was taken over by volunteers, members of the Surrey and Hampshire Canal Society, who worked for nineteen years (putting in millions of man hours) to restore it.

He said, 'Now they're cutting down trees to let the sunlight through. That's helped create a reed fringe that stops erosion. And to stop the water voles from being attacked by wild mink, along the canal there are mink rafts with traps on them.'

A silver Learjet sprang from the runway, howled overhead and went climbing off to the west, leaving behind odd surges and echoes of sound. Then he surprised me again. 'Did you know that of the forty species of dragonfly native to the UK, twenty-six live around Eelmore Flash? People throughout the country know about this spot, in dragonfly circles it's famous. You'll see emperors here, also willow emerald damselflies, migrant hawkers, ruddy darters and banded demoiselles, all just streaks of light among the reeds. *Beautiful.*'

Since he couldn't take off from the Flash, Cody hauled his water plane back to the shed, replaced the floats with wheels and got on with routine flight testing. Parliament, now preoccupied with the Zeppelin menace, worried about bombs being dropped on London. Cody shared his thoughts with a journalist. 'I would pin my faith on my own aeroplane with a 3 lb gun. I can fly higher and faster than any Zeppelin. Where would a Zeppelin be with me circling round it, firing little shots through every ballonet?'

He damned the War Office for its precipitate conversion to the idea that fixed-wing aircraft were better suited to aerial warfare than balloons and airships – then came up with a startling idea. 'Let me suggest we cannot wait for the government to make a navy in the air,' he said. 'It must be paid for by contributions from some of your rich men.' He went on to list them: 'Mr. Gordon Selfridge, Sir J. Lyons, Sir Thomas Lipton, people like Thomas Cook & Sons.' (This is a very modern notion; in an age in which PFIs, or private finance initiatives, have become fashionable we may yet see Kwik-Fit Eurofighters, Burger King Chinooks and Hercules transports with Carphone Warehouse written across their tails.) He then claimed he could build the very thing they needed. 'I have in mind an aeroplane such as the British government has never seen. The nation could be as secure in the air as they were when Nelson and his ships guarded the shores.' And, practical as ever, he had priced it down to the last strut. 'I reckon this aeroplane would cost me, in experimenting and materials, £8,000 before I delivered it.'

Something was clearly stirring, the ideas fizzing again, the old ambitions bubbling up. First, however, he had to get his water plane ready for the Coastal Britain contest. Yet, before the race had even begun, Lord Northcliffe announced a further prize: £10,000 for the first pilot to cross the Atlantic in either direction. Most thought the idea absurd – it would be another six years before Alcock and Brown succeeded – yet Cody immediately expressed interest.

Though he had intended to enter a monoplane in the Army Trials – before it was wrecked by a cow – the French and British military regarded monoplanes as inherently dangerous. (Then Blériot produced one with fortified top bracing wires which a Danish-born naturalized Briton named Gustav Hamel* flew upside down – he even looped the loop – at Brooklands.) Yet Cody, who believed there was no structural problem his genius couldn't overcome, not only chose the monoplane format for his trans-Atlantic bid, but decided to make it the biggest, heaviest aeroplane ever built, a vast double-decker with by far the longest wings – a hundred and twenty feet from tip to tip. He devised a fan-shaped tail and rounded rudder, a float beneath each wing, on the upper deck an enclosed two-berth cabin for the three-man crew, on the lower giant twin engines each producing in excess of 400 hp. Since no such aero engine existed he designed it himself, along with a four-bladed, chain-driven tractor propeller. Austro-Daimler, on receipt of a £600 deposit, agreed to do the work.

On Thursday 7 August 1913, he was due to fly his water plane to Calshot, the starting point on Southampton Water for the Coastal Britain race. There, having fitted his floats, Cody would, for the first time, take off from the sea. The race's chief organizer, who also

---

* A few months later the body of Hamel, a major Brooklands star, was pulled from the Channel by an Étaples herring boat. He remains of interest due to rumours, persisting for decades, that he was on his way to join the German air force.

happened to be Secretary of the Royal Aero Club, had just finished his own circumnavigation – a retired naval officer, he'd done it by boat – visiting each of the stopping places. Cody, thumbing through the maps, photographs and notes he'd prepared, spoke with his customary lack of modesty. 'I am going to be the man of the race. Everybody will be looking out for me.' The irritating thing, from the secretary's point of view, was that Cody was right. He admitted as much, saying that everywhere he had gone, people had asked, 'Will Cody get here?'

The August Bank Holiday fell three days before he was due at Calshot. On the Sunday he flew to Brooklands and, dressed in white, showed off his new machine to the crowd, then had tea at the Blue Bird as wide-eyed fans clustered around and snapped his photograph. On Thursday, before leaving for the coast, he took the famous ex-cricketer W. H. B. Evans (Oxford University, Gentlemen of England, sometime captain of Hampshire, now a colonial administrator on furlough from the Egyptian Civil Service) for a spin. It was a warm, windless morning, and Evans arrived early at the shed. The flight, initially, passed without incident. They arrived over Bramshot Golf Course, circled the clubhouse at 500 feet then turned back towards Laffan's Plain. But over Ball Hill, as Cody evidently began a low pass – he called it 'a zoom' – the plane seemed to buckle and break, throwing its occupants clear. Stunned observers saw Cody, distinguishable by his all-white flying suit, falling with arms spread wide as if trying to fly, to test certain notions of balance and control. The two men, when people reached them, lay only feet apart. (One of the first on the scene was young Frank Cody who, weeping inconsolably, crying, 'Dad, oh Dad, oh Dad,' threw himself across his father's body.)

Geoffrey de Havilland was still in bed when it happened. 'I heard the unmistakable note of Cody's open exhaust as he flew round Laffan's Plain. Suddenly the sound was cut off, as cleanly as if by a switch, and never restarted. Later I heard with great sorrow that this silence had signalled the end of his last flight.'

On 7 August 1913, Cody took the famous cricketer W. H. B. Evans,
for a spin. Over Farnborough's Ball Hill, attempting a low pass – he
called it a 'zoom' – the plane seemed to buckle and break.
Cody's son Frank was one of the first on the scene.

A wave of sorrow swept the nation. The *Daily Mail*, on its front
page, declared, 'The sad news of Colonel Cody's tragic death will
come as a painful shock to the British public which admired him
as a pioneer and loved him for his noble adventurousness of spirit
. . . The man-lifting kite was his device. He designed a practical
aeroplane before the Wrights had shown all the world how it should
be done. His courage was as indomitable as his good humour. He
combined the simplicity of a child with the heart of a hero. Success
came late to him but it came. Yet now, just a year after the victory of
his machine in the Army trials, he lies dead, and it is for us to lay the
wreath of honour on his bier.'

The King and Queen sent Lela a condolence telegram. ('It is
with great sadness . . . ') So did his old enemies at the War Office.
('The science of aeronautics in this country owes much to his mechanical genius, and courageous perseverance, and the War Office has
special reason to mourn the loss of his valuable service . . . ') Friends

at the brand-new Royal Flying Corps, Cody's 'Navy of the Air', would miss 'a clever and gallant man'; other colleagues felt that 'aviation and humanity have lost a most loveable and splendid character.' The *Aeroplane* carried a curious verse entitled 'Cody'.

> Crank of the crankiest, ridiculed, sneered at;
> Son of a boisterous, picturesque race,
> Butt for the ignorant, shoulder shrugged, jeered at;
> Flint-hard of purpose, smiling of face.
> Slogging along on the little-trod paths of life;
> Cowboy and trick-shot and airman in turn!
> Recklessly straining the quick-snapping lath of life,
> Eager its utmost resistance to learn.
> Honour him now, all ye dwarfs who belittle him!
> Now, 'tis writ large what in visions he read.
> Lay a white wreath where your ridicule riddled him,
> Honour him now, he's successful and – dead.

It was reprinted in the *Daily Mail*, and Aldershot's *Military Gazette*.

In a lecture to the Women's Aerial League the previous year Cody had said, 'I hope when my time comes that the death will be sharp and sudden; death from my own aeroplane, like poor Rolls . . .'

And then there was that extraordinary funeral with literally hundreds of wreaths: a decorous one from Buckingham Palace, from the Green Engine Company (who powered his Michelin Cup-winning machine) a four-bladed floral propeller measuring ten feet across, from 'the Hendon Aviators' a plane-shaped tribute with a card signed by each and dedicated to 'England's greatest airman'. (One of the signatories, who identified himself simply as 'A. Russell', would become better known as General Osipenko of the Russian air force.) And a wreath came, unexpectedly, from the London and Aldershot telephone operators 'in remembrance of a man who never lost his temper on the line'.

Cody's funeral has inspired comparisons with Princess Diana's.
At least 100,000 mourners lined the route from his home, while a
further 50,000 assembled in and around the cemetery. By order
of King George V the pipes and muffled drums of the
Black Watch accompanied the coffin.

A memorial fund was opened for Lela. Prince Arthur of
Connaught, Lord Salisbury, Winston Churchill and Lloyd George
organized, at the London Hippodrome, a celebrity matinee featuring leading theatrical luminaries such as Sarah Bernhardt, Gladys
Cooper and George Robey. The stars, who put on forty-five different acts between them, waived their fees, while the paying public
turned up in such numbers the last to arrive had to stand.

At the inquest a Gunner Maxted, who had been watching the
aeroplane through field glasses, said it had broken in two, the front
half dropping into a grove of trees near Ball Hill. Cody, he added,
had fallen first, followed by his passenger. He then contradicted the
evidence of a previous witness who – sending the assembled press
rushing off in search of telephones – claimed he had seen Evans suddenly reach out and seize Cody by the throat. Mervyn O'Gorman,

now Superintendent of the Royal Aircraft Factory, brought pro-
ceedings to a close when he said the accident had been caused by
the 'inherent structural weakness' of a wing spar which snapped
under stress. (Other expert witnesses, who claimed the spar had
failed when a propeller blade broke off and struck it, were ignored
by the coroner.)

Sam Cody was buried in Aldershot's Military Cemetery, on a hill
overlooking the spot where he made his first flight. Lela lies with
him. On their tombstone stands the figure of Christ with (for reasons
no one can remember) two fingers missing.

# Epilogue

'The Secret Factory?' said the woman seated in the central rotunda. She pointed. 'It's at the end of the passageway. Just push the door and go in.'

'Is there anything about Cody?' I asked.

She smiled. 'He's the first thing you'll see.'

It was true; a life-sized cardboard Cody cut-out confronted me as I entered the room. A replica of one of his kites hung from the ceiling, and a press-to-play film loop showed Army Aeroplane No. 1, in fuzzy black and white, bumping across the grass as a voice intoned, 'It's the 16th of October, 1908, and it was to be a momentous day for a tiny village named Farnborough.'

The new Business Park, set on the old 180-acre MoD site, was described by its present owners, Slough Estates, as 'a unique environment encompassing work, leisure, art and heritage' (also, they might have added, the expectation of large profits). The three-storey structure in which I stood was called the Hub. It had once, actually, been the 1938 Weapons Test Department, scheduled for demolition until some Slough operant noted its fine period features. Now fully restored, once again 'a striking art deco building', it contained, along with the Secret Factory, a cafe and eleven 'contemporary office suites.'

Slough Estates had, in fact, done some interesting things. For a start they'd found, scattered across the old Factory Site, the parts of a Cody-era Portable Airship Hangar designed to be put up or taken down by fifty men in just ten days. Now it had been re-erected in its original form. In 1915, when it last stood here, the steel stanchions, girders and lattice arches were covered with heavy-duty army canvas.

But today, left bare, and burnished till they shone, it became a huge, spidery, silvery sculpture (two hundred and fifty feet long, seventy feet high) which, frankly, looked stunning.

The Secret Factory was the term local people once used for the old MoD site. With its high security fences, locked gates, air of mystery and sense of self-containment – employees, coexisting quite happily, even had their own symphony orchestra – it belonged to a world few others understood. The room which had been given its name, though, was a disappointment – just a few models and photos, some written material, and a button labelled, 'Press to hear reminiscences of those who worked at this site years ago.' I listened, but none went back far enough for me. What I liked best were two photographs. The first showed one of Frank Whittle's earliest jet engines mounted on a tank and being used as a snow blower, the next a crowd of women from Building Q65, the Doping and Painting Workshop. (Since the dope contained toxic substances which entered their lungs, they needed careful supervision and plenty of lemonade – several pints daily being prescribed by War Office doctors as an antidote to the poisons.) Now *hundreds* posed for the camera, all huddled companionably close in their heavy bonnets and long coats.

A caption said that for the duration of World War One they were known as 'Cody's Girls' – this particular Cody, I suspected, being Sam's stepson, Vivian, who had managed Farnborough's fabric shop.

The Aviators' Café stood at the other end of the building. I entered a spacious, crowded room with comfortable furniture and soft blue ceiling lighting, bought a cheese sandwich and found a seat at a table for two. Its other occupant, a man with a long bony nose and a mane of white hair, said, 'Yes, yes,' rather impatiently when asked if I could join him. Yet he was happy enough to chat, and told me his father, a metallurgist, had worked here. 'I've just popped up from Brighton to see what they'd done to the place.' What he remembered from before had been people on bikes. 'At five o'clock every afternoon thousands would come charging out the gates. Often I'd ride down to meet my dad, and if you weren't careful you'd get flattened in the rush.' He smiled faintly. 'Possibly by some of the cleverest men in the land; the concentration of scientific talent here was remarkable. Flying talent, too. The Empire Test Pilots' School? Famous in its day, best in the world.'

'Top Gun,' I said.

'Exactly – that all started at Farnborough. So, of course, did the jet engine, the supersonic airliner, the vertical take-off fighter. Now, though, it's all been privatized, the boffins have moved to Boscombe Down.' He pulled a Slough Estates brochure towards him. 'It says they've planted 1,620 trees, and that £20 million is to be spent on a "Heritage Quarter". Well, that's good. At least an eye is to be kept on the past.'

I knew suddenly that I was finished with the past, that to all intents and purposes my book was done. I asked him to take some wine with me. Aviators' produced a nice Shiraz and, when I'd brought a couple of glasses to the table, he told me about his grand-father, a pilot with the Royal Flying Corps who'd been stationed at Stonehenge Aerodrome. 'It was owned by a pig farmer and known for its solstice gap – hangars placed to give views of the midsummer sunrise. The only problem was the stones themselves; they interfered with visibility, so some genius in the War Office proposed dropping enormous bombs on them.' He chuckled. 'Then my son became interested in flying – probably the old man's influence. While he was at school he won a place on one of these RAF Flying Scholarship Schemes. They sent him up to Scotland; but he got homesick, didn't enjoy it at all. Today's he's a dentist, making shedloads of . . . '

'Whereabouts in Scotland?' I asked.

'Oh, Lord. Let me see now. Dundee, I believe.'

'I think you'll find he stayed at small hotel opposite the Dundee Church for the Deaf. And the man who supervised his training was a chap named Lovat Fraser, a Wing Commander in the RAF Volunteer Reserve.'

'Really?' he said, giving me a rather odd look. 'Well, I'm actu-ally meeting him for lunch today. In Kingston. I'll ask.' He glanced at his watch. 'Blimey! Must dash. Thanks for the wine.' And then he was gone.

He'd had only a single sip, so I emptied his glass into mine and sat recalling the day, at least a decade earlier, I spent with Lovat Fraser. When, after getting airborne at Wick, I abandoned my attempt to traverse Britain in a series of flying lessons, I phoned Fraser to cancel the lesson booked with him. But he wouldn't take no for an answer, raved about Dundee, promised me a wonderful lunch and absolutely insisted I come.

We met in the crewroom of Tayside Aviation, the firm he had founded at Dundee Airport. A trim middle-aged man with a broad, open face and a frank manner, he'd won a contract, worth a million a year, to teach five hundred schoolkids to fly; a hundred he'd kept here, the rest were subcontracted out around the country. 'Some people think I'm a pukka RAF type,' he said engagingly, 'but actually I'm ex-motor trade; until they made me redundant I ran a British Leyland dealership. But I've always loved planes.' He pointed out the window at a small one with a boomlike fuselage and a high T-tail.

'That's yours. It's brand new, a Katana, from an Austrian firm which usually makes gliders. See the tail? It's a glider's tail; gliders come down in crops that would tear an orthodox tail right off. Brilliant! Any idea where you want to go?'

'Not really.'

'Well, it's all beautiful, stunning scenery, you'll see. Afterwards I'll take you to my brother's restaurant. By the way, you're going up with Phillip Hughes. He'll be here shortly.' He beamed. 'You haven't met Eileen!'

His wife, a blonde with a quick smile and glamorous good looks, ran the crewroom. Paying for my hour's lesson – £84.60 from brakes off to brakes on, inclusive of VAT and landing fees – I enquired about the weather. She said, 'You've got twenty-five kilometres visibility. There's cloud on the hills, and moderate ice in the cloud. Moderate turbulence, too.'

Idly perusing a map of Scotland on the wall I spotted a familiar name and suddenly had an idea. Phillip Hughes proved to be a brisk young man with brown hair and a firm chin. 'Any preference regarding our route?' he asked.

When I told him, he laughed.

We scrambled into the Katana's futuristic cockpit and closed its fighter-style canopy; this time I was not invited to take off, though once we had reached 2,000 feet, he said, 'You have control.'

But I hadn't. In this volatile little high-tech plane my hands and feet were activated by two sides of a brain that, for the moment, were no longer in touch. A bit of ragged Arctic cumulus floated over Dundee yet the Firth of Tay was irradiated by strong sunlight; as we crossed its dazzling epicentre the plane got a sharp little buffeting (solar winds!) that ceased as soon as the coast dropped behind. Now we headed west towards the area where once, for the *Observer*, I had walked the route taken by Macduff when, with ten thousand foot soldiers pretending to be trees, he marched off to confront Macbeth. Today, however, I planned to travel the eleven and a half miles from Birnam Wood to Dunsinane by air.

Shakespeare had probably heard the story while visiting Perth with a company of comedians in 1599. Now, banking over the pretty town of Birnam, I looked for the only oak – 'Let every soldier hew him down a bough, and bear't before him' – surviving from the original wood, and recalled the way its massive, half-fossilized branches had been propped on iron supports. Just across the Tay lay Dunkeld, which, with its extraordinary ruined cathedral, once held the primacy of Scotland. Queen Victoria liked Dunkeld. ('We stopped to let Vicky have some broth,' she wrote. 'Such a charming view from the window! Vicky stood and bowed to the people.') Beside a burn delineating the geological fault line dividing the Highlands from the Lowlands we saw the white house in which Beatrix Potter spent her holidays. We followed the Pass of Birnam and crossed the hill containing the present wood (hybrid larches 'planted' by a lazy Duke of Atholl who, having placed the seeds in hollow cannonballs designed to shatter on impact, fired salvoes of them to the top).

Outside the village of Murthly, last time, I had come upon a stern-looking old woman walking a Pekinese in a field of barley stubble. She told me Macbeth had been nothing more than a local psychopath. 'The Perth Lunatic Asylum used to be at Murthly,' she said. 'Had it been here then they'd have put him in a padded cell. Shakespeare dignified him in a way he didn't deserve.' Now, overflying Murthly, we spotted a skein of geese which, several thousand feet higher, were overflying us. In 1880 a local landowner, Sir William Drummond Stewart, imported a herd of buffalo and several Red Indians to tend them; after the Indians perished from pneumonia the buffalo roamed free, trampling crops, terrorizing the crofters and chasing travellers along the road.

Over Meikleour we glimpsed the world's highest hedge – a beech monster topping eighty-five feet – then a moment later saw the curve of the Tay, today a seductive Caribbean blue, which Macduff's army would have somehow crossed before heading for Dunsinane Hill. When I climbed it, hares had kept exploding out of the snow; now, though, it was hidden by banks of Eileen's moderately iced cloud. And I imagined again Macbeth hearing his messenger's incredulous words: 'I looked towards Birnam, and anon, me thought the wood began to move.' I now had the hang of the Katana and, carefully turning for home, told Phillip it was a true story – though, in fact, two years passed before Macduff finally put Macbeth to the sword. As he retrieved the controls and prepared to land back at Dundee I calculated that, cruising at eighty-five knots, we had gone from Birnam Wood to Dunsinane in a shade over seven minutes.

Back in the crewroom Eileen said, 'You're lunching with Lovat. Yes? Just take a seat, Alex. He won't be a moment.'

The only other person sitting there was a female flying scholar with crisp black curls and emerald-green eyes who managed to look sexy even in a flying suit one size too big. Waiting for her instructor, she told me she came from Ealing in West London, was doing A-levels (maths and physics) and wanted to fly Tornados for the RAF. 'Though,' she said cheerily, 'the idea terrifies my parents.' Yet it transpired that she wasn't the first in her family to fly, indeed could claim, on her mother's side, to be distantly related to a famous aviation pioneer. 'He was an author too. John William Dunne. Have you heard of him?'

'Of course.'

She looked immensely pleased. 'We've got his books.'

'Well, make sure you hang on to them. They're probably quite valuable.'

'My dad says he built his own aeroplane; it was kind of V-shaped, like a small stealth bomber, and could be flown hands-off.'

We were chatting about inherent stability – she knew absolutely everything about that – when her instructor walked in, plainly delighted that he'd got this gorgeous girl all to himself for the next hour. As she got up I wished her well; she smiled, thanked me and *shook my hand*.

I couldn't get over it – what class! (Good grip, too.) I still think of her from time to time, imagining John Willy's sense of utter wonderment had a Tornado gone rocketing past his window (specially with her at the controls).

I finished my wine then, standing outside waiting for the courtesy bus to Farnborough station, looked over to the Heritage Quarter where the Balloon Factory once stood. It had, as far as I could establish, been demolished in 1985, and I recalled being told that when they checked the building before the wreckers came, they found a small locked upstairs room that hadn't been entered for years. It contained a comfortable chair, a table, a kettle, a cup, a spoon, a jar that once probably held sugar and a tin with traces of dust that may have been tea leaves. The identity of the person who used that hideaway remains a mystery.

But not to me.

Cody's Tree.

# BIBLIOGRAPHY

*Early Aviation at Farnborough – The History of the Royal Aircraft Establishment* by Percy B. Walker C.B.E., Ph.D., F.R.Ae.S. Macdonald, London, 1971

*British Aviation – The Pioneer Years 1903–1914* by Harald Penrose. Putnam, London, 1967

*Sky Fever – the Autobiography of Sir Geoffrey de Havilland.* Airlife Publishing, Shrewsbury, 1979

*Pioneer of the Air – the Life and Times of Colonel S.F. Cody* by G. A. Broomfield. Gale & Polden, Aldershot, 1953

*The Flying Cowboy – Samuel Cody, Britain's First Airman* by Peter Reese. Tempus Publishing, Stroud, 2006

*The Chronic Inventor – the Life and Work of Hiram Stevens Maxim, 1840–1916* by James E. Hamilton. F.L.A. Libraries & Museums Department, Bexley, 1991

*The Air Pilot's Glossary & Reference Guide* by David Bruford. Airlife Publishing, Shrewsbury, 1994

*A Course in Elementary Meteorology.* Her Majesty's Stationery Office, London, 1978

*The Man Who Discovered Flight: George Cayley and the First Aeroplane* by Richard Dee. McClelland & Stewart, Toronto, 2007

'Aviation Pioneers: An Anthology', www.ctie.monash.edu.au/hargrave/pioneers.html

'The British Weather' by T.A. Harley, www.personal.dundee.ac.uk/~taharley/britweather.htm

# ACKNOWLEDGEMENTS

My thanks, firstly, to Charlotte Greig, senior editor at Picador, who not only managed to master, with astonishing speed, a subject as arcane as early British aviators, but then saw, better than I, what form it should take. During the years I worked on this book she was always positive and encouraging and, more to the point, she was always right.

Nicholas Blake, my copy editor, gave his usual master class on the English language, and I thank him. (Having Nick scrutinize one's prose is like experiencing a particularly rigorous tutorial by email.) My thanks also to Picador's designer, Stuart Wilson, who produced such memorable covers for both the hardback and paperback versions.

My initial access to many of the personalities and events described in these pages was through Harald Penrose's magnificent *British Aviation – The Pioneer Years 1903–1914*. Penrose, himself a famous pilot, knew most of the people he wrote about with such authority, style, insight and wit, and I'm grateful to his son, Mr Ian Penrose, for permission to quote from this important book.

I found Sir Geoffrey de Havilland's autobiography, *Sky Fever*, invaluable, and recommend it to anyone wishing to know more about that remarkable man. I'm grateful to Mr Philip Birtles, also to Mr Ralph Steiner of the de Havilland Aircraft Heritage Centre, for permission to use photographs from the book.

Brian Riddle, Librarian of the Royal Aeronautical Society, not only allowed me to quote from the journals of Thomas Walker, Francis Wenham, John Stringfellow, Percy Pilcher and Sir George Cayley, but also gave me access to the Society's archive of pictures and photographs. His guidance through that magnificent collection was invaluable.

I have already written about Jean Roberts, so here I will merely note that Sam Cody is fortunate to have a guardian so utterly committed to his memory.

I have taken material from T. A. Harley's fascinating website on British weather (his fans will also know him as Professor of Cognitive Psychology at Dundee University) and thank him for allowing me to do so.

Ali Fujino of the Drachen Foundation allowed me to use a Cody kite picture, and a quotation from Cody's diary.

I am grateful to Sharon and Terry Munns, owners of Mussel Manor, Shellbeach, for making me welcome when I blundered, uninvited, onto their property. Also to Diana Campbell for alerting me to the existence of the Cody Gallery at the Aldershot Military Museum, and her daughter Sophie, one of the country's finest travel writers, for urging me to visit Brooklands.

My own choice of pioneers may be viewed as eccentric. Why, for example, is there no mention of Sir Frederick Handley Page? He created such legendary aeroplanes as the HP 42 Hannibal (for Imperial Airways), the World War Two Halifax bomber and, for Britain's Cold War air defences, the Victor four-engine jet. Also, like de Havilland and Verdon Roe, he went on to found a great company.

Yet he had no link with the Balloon Factory – or none I could discover. Neither, of course, did Verdon Roe, but he built his first aircraft at Brooklands, just a hop and a skip from Farnborough. Cody, who dropped in constantly, certainly knew Roe; indeed, it was at Brooklands that Roe, quite possibly, beat him into the air, thus becoming the first man in Britain to fly, and a key part of this saga.

At a party thrown by British Airways in 1986 to mark the fiftieth anniversary of its Hong Kong service (operated by a tiny de Havilland biplane) I met an elderly American who'd been an aviation trailblazer out East. I can't recall his name, but vaguely remember him claiming that sky-writing was a sure-fire way of getting girls in China. Now, years later, I suddenly find myself wondering if a sky-writing story (nothing to do with sex) I attributed to 'Rupert Wood' – the octogenarian in my Prologue – actually came from him. I only mention it in case any of the American's descendants should read this book and think, 'Hey! Wasn't it my granddaddy/great-granddaddy actually did that?'

Finally I should like to thank my family – with love – for their patience and eternally good-humoured support.

PICTURE ACKNOWLEDGEMENTS

Brompton Hall Brochure: pp. 41, 54

Courtesy de Havilland: pp. 116, 148

Drachen Foundation: p. 169

Fleet Air Arm Museum: p. 165

Imperial War Museum: p. 182 (negative number RAE-0 412)

From the collection of Leonard Lewis: p. 63

*Mirror of Life Magazine*: p. 159

The *Navy and Army Illustrated*, 21 January 1898: p. 58

From the collection of Jean Roberts: pp. 161, 172, 175, 195, 203, 215,
219, 223, 229, 231, 240

Royal Aeronautical Society Library: pp. 15, 23, 29, 30, 33, 36, 38, 56,
67, 79, 83, 88, 101, 104, 107, 127, 129, 143, 197, 207

Science Museum/Science and Society Picture Library:
p. 90 (number 10462293 'Sir Hiram Maxim')